THE ROLE OF INTELLECTUAL PROPERTY RIGHTS IN AGRICULTURE AND ALLIED SCIENCES

THE ROLE OF INTELLECTUAL PROPERTY RIGHTS IN AGRICULTURE AND ALLIED SCIENCES

Edited by

Chandan Roy, PhD

Apple Academic Press Inc.
3333 Mistwell Crescent
Oakville, ON L6L 0A2
Canada

Apple Academic Press Inc.
9 Spinnaker Way
Waretown, NJ 08758
USA

First issued in paperback 2021

Exclusive worldwide distribution by CRC Press, a member of Taylor & Francis Group

ISBN 13: 978-1-77463-392-2 (pbk)
ISBN 13: 978-1-77188-698-7 (hbk)

Library and Archives Canada Cataloguing in Publication

The role of intellectual property rights in agriculture and allied sciences / edited by Chandan Roy, PhD.

Includes bibliographical references and index.
Issued in print and electronic formats.
ISBN 978-1-77188-698-7 (hardcover).--ISBN 978-1-351-12528-4 (PDF)

1. Agricultural biotechnology--Patents. 2. Intellectual property. I. Roy, Chandan, 1983-, editor

S494.5.B563R65 2018 631.5'233 C2018-903477-7 C2018-903478-5

CIP data on file with US Library of Congress

Apple Academic Press also publishes its books in a variety of electronic formats. Some content that appears in print may not be available in electronic format. For information about Apple Academic Press products, visit our website at **www.appleacademicpress.com** and the CRC Press website at **www.crcpress.com**

DEDICATION

This book is dedicated

to

my beloved daughter Ahana

CONTENTS

ABOUT THE EDITOR

Chandan Roy, PhD
Assistant Professor-cum-Junior Scientist,
Department of Plant Breeding and Genetics,
Bihar Agricultural University, Sabour, Bhagalpur, India

Chandan Roy, PhD, is an Assistant Professor-cum-Junior Scientist in the Department of Plant Breeding and Genetics, Bihar Agricultural University, Sabour, Bhagalpur, India. He is the author of three books, two book chapters, more than 13 peer-reviewed journal research papers, and six articles, and he has presented research findings at more than 10 seminar/symposia. Presently, he is the In-Charge of Intellectual Property Right Cell of Bihar Agricultural University, Sabour, as well as the university's Nodal Officer of the national-level program on PPV&FRA-2001 (Protection of Plant Varieties and Farmers' Rights Act, 2001). He has delivered lectures on different aspects of international property rights (IPR) at state- and national-level institutes. He has also conducted a national conference on IPR in agriculture. In addition, he is engaged in an awareness generation program among the farmers of Bihar, India, regarding farmers' rights; he has trained more than 3500 farmers, students, and scientists regarding IPR and farmers' rights. Dr. Roy is a member of the national-level expert committee on registration of essentially derived varieties (EDVs) under the PPV&FR Act 2001, Government of India. He is also a member of Project, Monitoring and Evaluation (PME) Cell of the university extension and training group. Dr. Roy is also engaged in teaching of undergraduate and postgraduate students and directs research activities in the wheat and cauliflower improvement program.

LIST OF CONTRIBUTORS

Stephen Amoah
Council for Scientific and Industrial Research – Crops Research Institute, P.O. Box 3785, Kumasi, Ghana

P. K. Bhattacharya
Bidhan Chandra Krishi Viswavidyalaya, Mohanpur, Nadia, West Bengal, India

R. C. Chaudhary
Chairman, Participatory Rural Development Foundation (PRDF), Gorakhpur (U.P.) 273014, India

Arpita Das
Department of Genetics and Plant Breeding, Bidhan Chandra Krishi Viswavidyalaya, Mohanpur, Nadia, West Bengal, India

Arvind Gupta
ICAR-National Bureau of Agriculturally Important Microorganisms, Kushmaur, Mau Nath Bhanjan 275103, Uttar Pradesh, India

Sanjay Kumar Gupta
ICAR-National Bureau of Agriculturally Important Microorganisms, Kushmaur, Mau Nath Bhanjan 275103, Uttar Pradesh, India

Ajeet Kumar
Regional Research Station, Madhopur, West Champaran–845454, Dr. Rajendra Prasad Central Agricultural University, Bihar, Pusa (Samastipur), India

Arun Kumar
Department of Seed Science and Technology, Bihar Agricultural University, Sabour–813210, India

Hardesh Kumar
ICAR-National Bureau of Agriculturally Important Microorganisms, Kushmaur, Mau Nath Bhanjan 275103, Uttar Pradesh, India

Jitesh Kumar
Department of Molecular Biology and Genetic Engineering, Bihar Agricultural University, Sabour–813210, India

Mahesh Kumar
Department of Molecular Biology and Genetic Engineering, Bihar Agricultural University, Sabour–813210, India

Mukesh Kumar
Department of Seed Science and Technology, Dr. Kalam Agricultural College, Bihar Agricultural University, Kishanganj, Bihar – 855107, India

Ravi Ranjan Kumar
Department of Molecular Biology and Genetic Engineering, Bihar Agricultural University, Sabour–813210, India

Santosh Kumar
Department of Plant Pathology, Bihar Agricultural University, Sabour–813210, India

Sujit Kumar
U.P. Council of Agricultural Research, Lucknow – 226 010 (U.P.), India

Tribhuwan Kumar
Department of Molecular Biology and Genetic Engineering, Bihar Agriculture College, (BAU), Sabour–813210, India

Ujjwal Kumar
Department of Plant Breeding and Genetics, Bihar Agriculture College (BAU), Sabour–813210, India

Aniruddha Maity
Division of Seed Technology, ICAR- Indian Grassland and Fodder Research Institute, Jhansi, U.P. – 284003, India

Hillary Mireku Bortey
Council for Scientific and Industrial Research – Crops Research Institute, P.O. Box 3785, Kumasi, Ghana

Anirban Mukherjee
Social Science Section, ICAR-Vivekananda Parvatiya Krishi Anusandhan Sansthan, Uttarakhand– 263601 & Division of Agricultural Extension, ICAR-Indian Agricultural Research Institute, New Delhi–110012, India

Arindam Nag
Dr. Kalam Agricultural College, Bihar Agricultural University, Kishanganj, Bihar – 855107, India

Abhishek Parashar
ICAR-National Bureau of Agriculturally Important Microorganisms, Kushmaur, Mau Nath Bhanjan 275103, Uttar Pradesh, India

Ganesh Patil
Vidya Pratisthan's College of Agriculture Biotechnology, Vidyanagari, Baramati–413133, India

Chandra S. Prabhaker
Department of Entomology, Bihar Agriculture College (BAU), Sabour–813210, India

Kumari Rajani
Department of Seed Science and Technology, Bihar Agricultural University, Sabour–813210, India

Tushar Ranjan
Department of Molecular Biology and Genetic Engineering, Bihar Agricultural University, Sabour–813210, India

Renu
ICAR-National Bureau of Agriculturally Important Microorganisms, Kushmaur, Mau Nath Bhanjan 275103, Uttar Pradesh, India

Anita Roy
Bidhan Chandra Krishi Viswavidyalaya, Mohanpur, Nadia, West Bengal, India

Chandan Roy
Department of Plant Breeding and Genetics, Bihar Agriculture College (BAU), Sabour–813210, India

Koushik Roy
RRS (Hill Zone), Uttar Banga Krishi Viswavidyalaya, Kalimpong, Darjeeling, West Bengal–734301, India

Bholanath Saha
Dr. Kalam Agricultural College, Bihar Agricultural University, Kishanganj, Bihar, India

Pramod Kumar Sahu
ICAR-National Bureau of Agriculturally Important Microorganisms, Kushmaur, Mau Nath Bhanjan 275103, Uttar Pradesh, INDIA

Upasana Sahu
ICAR-National Bureau of Agriculturally Important Microorganisms, Kushmaur, Mau Nath Bhanjan 275103, Uttar Pradesh, India

Khan Mohammad Sarim
ICAR-National Bureau of Agriculturally Important Microorganisms, Kushmaur, Mau Nath Bhanjan 275103, Uttar Pradesh, India

Anjan Sen
Patent Attorney & IPR Advocate Anjan Sen & Associates, 17,Chakraberia Road South, Kolkata–700025, India

Pawan Kumar Sharma
ICAR-National Bureau of Agriculturally Important Microorganisms, Kushmaur, Mau Nath Bhanjan 275103, Uttar Pradesh, India

Kumari Shubha
Germplasm Evaluation Division, ICAR-National Bureau of Plant Genetic Resources, New Delhi–110012, India

Prabhash K Singh
Department of Plant Breeding and Genetics, Bihar Agriculture College (BAU), Sabour–813210, India

Ravi S. Singh
Department of Plant Breeding and Genetics, Bihar Agriculture College (BAU), Sabour–813210, India

Seweta Srivastava
School of Agriculture, Lovely Professional University, Phagwara – 144 411, Punjab, India

LIST OF ABBREVIATIONS

ACTS	African Center for Technology Studies
AO	Appelation of origin
BAU	Bihar Agriculture University
BMC	Biodiversity Management Committees
CAGR	compounded annual growth rate
CB	claw back
CBD	convention on biological diversity
CML	chronic myeloid leukemia
COP	Conference of Parties
CSIR	Centre of Scientific and Industrial Research
CSP	cold shock protein
DMCA	Digital Millennium Copyright Act
DUS	distinctiveness, uniformity and stability
EDV	essentially derived variety
EMR	Exclusive Marketing Rights
EPC	European Patent Convention
EPO	European patent office
EU	European Union
FAO	Food and Agricultural Organization
FR	farmers' right
FV	farmers' variety
GATT	General Agreement on Tariffs and Trade
GI	geographical indication
GMS	genetic male sterile systems
IAS	invasive alien species
ICAR	Indian Council of Agricultural Research
IDA	International Depositary Authority
ILO	International Labor Organization
IP	intellectual property
IPAB	Intellectual Property Appellate Board
IPO	Indian Patent Office
IPR	intellectual property rights

IRRI	International Rice Research Institute in the Philippines
ITK	indigenous traditional knowledge
JPO	Japanese patent office
KP	Keekar pod extract
MCC	microbial culture collection
MTCC	microbial type culture collection
NARS	National Agricultural Research Systems
NBA	National Biodiversity Authority of India
NCCS	National Center for Cell Science
NCI	National Cancer Institute
NGB	National Gene Bank
NKP	neem kernel extract
PBR	people's biodiversity register
PBR	plant breeder's rights
PRDF	Participatory Rural Development Foundation
PVP	plant variety protection
SARS	severe acute respiratory syndrome
SAU	State Agriculture Universities
SBB	State Biodiversity Boards
SMTA	standard material transfer agreement
SPS	sanitary and phyto-sanitary measures
SWOT	strengths, weaknesses, opportunities and threats
TBT	technical barriers to trade
TK	traditional knowledge
TKDL	traditional knowledge digital library
TM	trademark
UN	United Nations
UNEP	United Nations Environment Program
UPOV	Union for the Protection of New Varieties of Plants
USDA	US Department of Agriculture
WCED	World Commission on Environment and Development
WCT	WIPO Copyright Treaty
WFS	World Food Summit
WHO	World Health Organization
WIPO	World Intellectual Property Organization
WTO	World Trade Organization

ACKNOWLEDGMENTS

I wish and express my sincere thanks and gratitude to all the contributors for spending valuable time to express their ideas and experiences in the field of intellectual property rights in the form of a book chapter. I express my gratitude to Dr. R. R. Hanchinal, Chairperson, PPV&FRA, New Delhi; Dr. Sudhir Kochhar, Agricultural Biodiversity and Intellectual Property Expert and Former National Coordinator, NAIP, ICAR, New Delhi, Dr. U. K. Dubey, Regional Deputy Registrar, PPV&FRA, New Delhi for time-to-time guidance and suggestions on various aspects of IPR. I am also thankful to Bihar Agricultural University for providing me an opportunity to work with the IPR Cell of the University; this experience has helped me to formulate the idea of this book. I extend my sincere thanks to all who are directly or indirectly related with the work.

PREFACE

Currently the technology innovation in any field of research is facing a major challenge due to piracy and increased globalization. As per the TRIPS agreement under the World Trade Organization (WTO), it has become mandatory for all the member countries to maintain minimum standards of their intellectual properties and their protection through IP laws. Providing suitable IP laws in any field of technology can help to accelerate industrial growth, produce quality products, and ensure sustainable use. It acts as a magnet for foreign investment, encourages innovation, and uplifts the economic status of the nation. From ancient times, ensuring IP rights by inventors is one of the oldest measures to protect their inventions. Among the earliest document of a patent is of "some kind of newfangled loaf of bread" back in 600 BC. This shows that inventors realized the importance of IPR long ago and saw the need to protect intellectual property through legal measures.

Since beginning, agriculture has been the back bone of every nation for social and economic development. Presently, most of the developing countries are dependent on agriculture for economic growth and food security; a significant amount of gross development product (GDP) comes from the agricultural sector. Plant genetic resources are considered as the heritage of mankind and was not included as patentable subject matter in most of the countries in the world. However, with the development of the TRIPS agreement under the WTO, there is a measure to protect plant genetic resources of the respective member state by IP laws. With the advent of the WTO, IP laws in the field of agriculture and agriculture biotechnology have been broadened to protect many other agricultural properties that could not be protected under patent or copyrights, the two oldest form of IP laws. Currently, microbes, plants, animals, and their parts and components are under the jurisdiction of patentability in many countries.

Almost all types of IP laws are applicable directly or indirectly to agricultural technology. However, this book focuses on the IP laws in detail that are directly related to agricultural technology that involves mostly microbes, plants, genes, or their parts. The first chapter deals with an introduction to different IP laws in brief and the usefulness of patents in agriculture in detail. The second and third chapters cover the importance of biodiversity and their conservation in developing countries, such as in India, and their sustainable utilization and benefit sharing respectively.

As an ancient science, agriculture is rich in traditional knowledge associated with communities and local regions of different countries. Therefore, promotion of such traditional knowledge for the benefit of society through legal protection has enormous opportunities. Chapter 4 of this book discusses protection of traditional knowledge through legal means and their promotion for social benefits. Development of new plant varieties is a continuous process for crop improvement and food security. Therefore, protection of Plant Breeders Rights was started since the development of the International Union for the Protection of New Varieties of Plants (UPOV) in 1961. Before the establishment of the WTO in 1994, many countries were not part of UPOV but the Trade-Related Aspects of Intellectual Property Rights (TRIPS) agreement compelled all member countries to adopt plant breeder's rights. Chapter 5 in this volume covers the development on IP regulations related to the protection of Breeder's Rights. Similarly, protection of plant varieties from unauthorized use through legal means is equally necessary as protecting breeder's rights. Most countries have adopted the use of patents to protect their plant varieties. However, a unique act was developed in India where farmers' rights were given prime importance, including breeders' right and researchers' rights. This unique act is now being adopted by many other countries in the world.

In Chapter 6, the author deals with the Protection of Plant Varieties and Farmer's Rights act in general and the Farmers' Rights in particular. In Chapter 7, protection of agriculturally important microbial genetic resources through IP laws is discussed. Geographical indication (GI) of goods is getting huge attention because several international controversies have been discussed between countries to resolve many issues on GI. One of the popular issue is GI of basmati rice, the most popular quality rice grown mostly in India and Pakistan. In the eighth and ninth chapters, the authors discuss GI of goods in relation to India and global needs.I hope the book has covered major aspects of IP laws that are directly applicable in the field of agriculture. This book will be beneficial for agricultural students, teachers, researchers, stockholders to understand IPR and advancement made in respect to different IP laws. I also invite readers for their valuable suggestions, inputs on subject matter presented in this book for further improvement.

Agricultural biotechnology is now an established field of agricultural science. IP rights in biotechnology have started a new path, as with the case of Diamond vs. Chakraborty in the year 1980. Patents were being granted on naturally occurring genes, cDNA, RNAs, or proteins if it fulfills the criteria of patentability. This is why understanding the fact of patenting in biotechnology as a whole and as gene in particular is very important. Chapter 10 deals with the patent of genes and other products of biotechnology. In the last chapter the author discusses the awareness generation in the field of agriculture regarding IP rights on various components with the brief introduction of different IP laws.

I hope the book has successfully covered the major aspects of IP laws that are directly applicable in the field of agriculture. This book will be beneficial for agricultural students, teachers, researchers, and stockholders to understand IPR and advancements made in respect to different IP laws.

—Chandan Roy, PhD

CHAPTER 1

IPR IN AGRICULTURE AND THEIR PROSPECTS

ANJAN SEN

Patent Attorney and IPR Advocate, Anjan Sen and Associates, 17, Chakraberia Road South, Kolkata–700025, India, E-mail: anjanonline@vsnl.net; info@ipindiaasa.com

CONTENTS

1.1 WHAT ARE INTELLECTUAL PROPERTY RIGHTS (IPR)?

Intellectual property (IP) as the name suggests refers proprietary rights on *creations of the human mind or intellect.* Development of intellectual creations from abstract ideas to valuable expressions/practical applications, which may be in the field of science, arts, or literature or any other forms of original creations involves intellectual efforts, time, and financial investment and hence calls for exclusivity in favor of the creator/author. Thus, such creators/authors or their legal assigns usually seek a return on the efforts and investments made by way of claiming the same as proprietary intellectual property rights (IPRs) on such creations of the human intellect. Importantly, the origin of all IP is always attached to the human creativity and not to any legal entity and evidences the power of the human mind/intellect in generating intellectual property rights. The IPRs in the field of agriculture is usually categorized as industrial property rights including Patent rights, Design copyright, Trademark, Plant varieties rights, all being relatable to industrial utility of such property rights. All such forms of IPRs entitle the proprietor to certain exclusive rights and privileges for a given period of time in favor of the proprietor including creator/inventor/author or his/her legal heirs/successors or assigns.

1.2 THE NEED FOR IPR IN AGRICULTURE

Agricultural science is a multidisciplinary stream that involves study of the science and management of nature and biological systems for the sustainable production of food and fiber involving cultivation and breeding of animals, plants, and fungi for food, fiber, biofuel, medicinal plants, and other products. With the advancements in technology, there has been a revolution in all fields of agriculture spread over several decades now, especially over the last decade, primarily in the area of agro-biotechnology. Advancement in technology usually leads to the development of new products and/or processes and related intellectual creations and possibly involve further intellectual efforts in introducing the same in the related trade and business or industrial purposes, which in turn leads to further generation of other forms of intellectual creations, all of which are valuable to the inventor/creator/author of such intellectual creations and also to their legal heirs/successors. Thus, there arises the need to safeguard and protect such creative efforts as IP free from any unauthorized use or dilution and the related laws to protect and safeguard such IPRs.

Broadly, advancements in the field of agricultural science could include advancements in agro-mechanical, agro-chemistry, and agro-biotechnology, which attract different forms of IPR: patents for new inventions such as technical advancement in any agricultural product (constitution/characteristics) and/or process of manufacture; trademarks involving brands/marks/devices/signs under which the agricultural product or process or related services originating from a defined source is introduced and marketed in the related trade and business; industrial designs for covering any new and original shape, configuration, surface pattern, or ornamentation or color combinations applied to any agricultural equipment/device/article; geographical indications (GI) that are special characteristics attached to any agricultural produce of any particular geographic region; and copyrights on any original literary work or artwork in any product brochure/package content and plant variety rights for protecting new plant varieties.

1.3 THE LEGAL AND STATUTORY RIGHTS AND REQUIREMENTS TO QUALIFY SUBJECTS UNDER VARIOUS FORMS OF INTELLECTUAL PROPERTY RIGHTS RELEVANT TO AGRICULTURAL SCIENCES

While the above discussed forms of IPRs did exist since long and have been recognized as valuable intellectual creations, the extent of protection and enforcement of IPRs varied widely from country to country as such rights are country specific and governed by the laws of the land. However, as the importance of IPRs grew, the need for a common basic minimum form of IPR protection in all jurisdictions and reciprocal rights in various jurisdictions became important, and the nations started to debate on how differences in scope and coverage in various jurisdictions raised complications in international economic relations. In order to introduce uniformity in IPR including to settle the disputes more systematically, the need for formulation of new internationally agreed trade rules for IPRs was greatly felt. This resulted in the much needed platform for discussions on bringing uniformity in IPR laws in various jurisdictions and the well-known and globally popular *Trade-Related Aspects of Intellectual Property Rights (TRIPS) Agreement* and contributions in setting standards of IPRs and need for its legal protection.

1.4 THE TRIPS AGREEMENT AND STREAMLINING OF IPR LAWS IN VARIOUS MEMBER COUNTRIES

The international agreement on TRIPS is a common agreement agreed upon by all the member nations of the World Trade Organization (WTO) to narrow the gaps on basic minimum standards of IPRs and to bring the IPRs under a common internationally agreed framework. The TRIPS Agreement thus prescribed minimum levels of legal protection that each member government/country has to enact to grant legally enforceable IPRs in member countries and thereby maintain a balance between the long-term benefits and possible short-term costs to society. Thus, the scope, purpose, and objectives of the TRIPS Agreement can be said to have laid the basic foundation of global recognition of basic forms and scope and legal coverage of IPR to be guaranteed by the member countries.

The developments of the IPR laws and related rights in the above backdrop of the TRIPS Agreement now forms basis of IP laws covering various forms of IPRs among the members of the TRIPS Agreement and WTO member countries to which India along with other member countries including developed states/regions like the USA and European Union members are signatory. Relevant forms of such IPRs of interest that attract the agricultural sector are discussed below.

1.5 DIFFERENT FORMS OF IPRs AND THEIR RELEVANCE TO THE AGRICULTURAL SECTOR

1.5.1 TRADEMARK

A "trademark" as a form of IPR is basically an identity of goods and services offered in the trade and business and its sources/origin to benefit the customer/consumer and usually include distinguishable identity as a word, signature, device, letter, numeral, brand, heading, label, stylized artwork, and even color combinations and shape of goods (subject to not being a shape solely dictated by the function or character of the goods being sold under that shape) or any combination thereof or any other distinguishing identity of proprietary value.

For the purpose of registration as an IPR with exclusive privileges in favor of its owner/proprietor, a trademark chosen should be capable of distinguishing goods or services of one person from those of the others. Further, it should not be deceptively similar to an existing mark of another person apart from being not expressly prohibited under any law. Marks devoid of any distinctive character or that are only indicative of the kind, quality, quantity, purpose, value, or geographical origin of the goods, or that are marks already in vogue in the trade due to their customary use may not be registered.

Registration of a trademark provides for exclusive rights in favor of the proprietor to use and apply the registered mark in relation to the goods and services for which the same is registered for a period of 10 years, which can be renewed thereafter for each subsequent term of 10 years and continued to be maintained.

Examples of some exemplary popular trademarks as IPRs relevant to agriculture industry and related trade and business include in respect of product tea brands/identity such as Brooke Bond®, Taj Mahal®, Tetley®, Lipton®, etc., or leading noodle brands in India such as Maggi®, a product of Nestle; Sunfeast Yippee!®, which belongs to the Indian conglomerate ITC; and Top Ramen®, product of the Japanese company Nissin, ,etc.

In India, the law of trademarks is currently governed by The Trademarks Act, 1999 (www.ipindia.nic.in/acts-rules-tm.htm).

1.5.2 INDUSTRIAL DESIGNS

A design as a form of IPR refers to the new or original features of shape, configuration, pattern, ornamentation, or composition of lines or colors applied to any article, whether in two- or three-dimensional (or both) forms applied by any industrial process or means (manual, mechanical, or chemical) separately or by a combined process, which in the finished article appeals to and judged solely by the eye. Design does not include any mode or principle of construction or anything that is a mere mechanical device.

The registration of a design confers upon the registered proprietor the exclusive right to apply new or original design to the article in the class in which the design has been registered. Registration initially confers this right for 10 years from the date of registration. It is renewable for a further period of 5 years.

Some exemplary design registration applicable to agricultural products/articles includes new or original designs of "C"-shaped fertilizer tablet, sugar cubes, and even new or original designs of ploughing/agricultural implements, which apart from serving the functional part also have distinctive eye appeal. In India, the law of industrial designs is currently governed by The Designs Act, 2000 (www.ipindia.nic.in/designs.htm).

1.5.3 GEOGRAPHICAL INDICATIONS

Geographical indicator (GI) is yet another form of IPR attached to goods originating from a particular geographic region and the rights are based on

(a) an indication;
(b) originating from a definite geographical territory;
(c) used to identify agricultural, natural, or manufactured goods produced or processed or prepared in that territory; and
(d) having a special quality or reputation or other characteristics.

Importantly, while trademark and GIs both are relevant to apply distinguishing indications for goods, one basic difference between trademarks and GIs rights is that while trademarks are allowable as private rights, the GI rights are not private rights but can be claimed on behalf of the association/group protecting the interests of the growers or producers of goods attaching such GI based on the produce originating in certain geographic region only.

In India, the GIs as a form of IPR is currently governed by Geographical Indications of Goods (Registration and Protection) Act, 1999 (www.ipindia.nic.in/gi.htm). Any registered GI tag on a product ensured that none other than those registered as authorized users (or at least those residing inside the geographic territory and involved in such manufacture) are allowed to use the popular product name/GI attached to the product of such origin. Among several GIs of agricultural goods registered in India, some of the famous GI-protected goods are discussed below:

Darjeeling Tea (www.teaboard.gov.in/pdf/policy/geographical_indication_for_darjeeling_ tea.doc), a rare coveted brew consumed globally, is only grown in India and is protected as a registered GI. Considering that such characteristics of Darjeeling Tea would be attached to all tea grown in Darjeeling, only the original growers of Darjeeling Tea are authorized to use the GI "DARJEELING" Tea in respect of their produce in Darjeeling, India. Thus, to safeguard the importance of Darjeeling and tea grown there, especially the original growers of Darjeeling Tea, the Tea Board of India has obtained registration of Darjeeling Tea as GI under the GI Act in India. The Tea Board as a custodian of the interest of the tea growers in Darjeeling involved in growing exclusive Darjeeling Tea with special characteristics and attributes, which is specially attached to tea grown in Darjeeling, West Bengal, India, has protected and preserved the word DARJEELING in respect of the goods/product Tea and also devised a logo to be authorized in packages of tea of original growers of Darjeeling Tea.

Application Details (http://ipindiaservices.gov.in/GirPublic/ViewApplicationDetails)

According to Indian GI Registry records, the "Darjeeling tea" (word) is registered under GI No. 1 dated October 27, 2003 and "Darjeeling Tea logo" under GI No. 2 dated October 27, 2003, in the name of Tea Board, India.

Thus, only duly authorized growers of Darjeeling Tea would be allowed to use the GI DARJEELING word and logo in respect of tea grown and produced in Darjeeling, West Bengal, India, and no one else.

Tea Board of India while obtaining registration of DARJEELING as a GI has specified Darjeeling tea by some specific characteristics and covers all stages from the production level to the export stage and meets the dual objective of ensuring that (a) tea sold as Darjeeling tea in India and worldwide is genuine Darjeeling tea produced in the defined regions of the District of Darjeeling and meets the criteria laid down by the Tea Board and (b) all sellers of genuine Darjeeling tea are duly licensed. This licensing program affords the Tea Board the necessary information and control over the Darjeeling tea industry to ensure that tea sold under the GI/certification marks adheres to the standards for DARJEELING tea as set forth by the Tea Board.

Importantly, while under the TRIPS definition of "Goods" for the purposes of GIs that refer to goods in general, the Indian GI Act specifies the goods to be either agricultural goods or natural goods or manufactured goods that can qualify as a GI. Further, under the Indian GI Act, if a producer applies for a GI for a manufactured good, it is important to ensure that at least one of the activities of either the production or the processing and preparation of the manufactured good must take place in the territory to which the goods are attached as a GI.

This can be well appreciated in the case of "Darjeeling tea." Darjeeling tea involves processing of green tea leaves plucked from the tea bushes through a range of processing steps before turning into the final product (called "manufactured tea"), which is ultimately traded in the market. Hence, "Darjeeling Tea" being a product of manufacture only if the tea leaves are plucked from the Darjeeling region; the GI Act will not statutorily allow the final product to be designated as Darjeeling tea, unless the processing of the tea also takes place within the Darjeeling region. Such requisites attached to GI of manufactured goods thus further ensure standards of processing linked to the geographic location in relation to the end product carrying such GI.

Coorg Orange (http://ipindiaservices.gov.in/GirPublic/ViewApplicationDetails), also called Coorg mandarin, is also a registered GI in India, which is a cultivar of orange from Kodagu in Karnataka, India. Coorg oranges are regarded as man-made hybrids of mandarins (*Citrus reticulata*). Greenish-yellow in color, they have a tight skin and a sweet-sour taste, which differentiates them from the Nagpur oranges. It was given the GI status on March 31, 2005 with GI No: 33 in the name of Dept. of Horticulture, Govt. of Karnataka & Biotechnology Centre Hulimavu, Bangalore, Karnataka.

Basmati Rice (Basmati Case Study: www1.american.edu/TED/basmati.htm; Basmati Patent US5663484 -, Rice Tec in) is a top-quality rice from the Punjab provinces of India. The word means "fragrant earth," and the rice is a slender aromatic and long grain variety that originated in this region and is a major export crop. In 1997, the US Patent Office had granted a patent to US firm Rice Tec Co. (patent number 5663484) for a variety called Texmati rice (20 claims). Texmati rice patent application initially had 16 claims similar to the well-known characteristics of Indian Basmati rice. According to GI, Basmati being a well-known Indian rice variety, which could not be patented as such in any other regions, is a natural produce with its well-known and publicly available inherent characteristics. According to Indian GI registry, Basmati is registered with GI No: 145 dated February 15, 2016, in the name of "The Agricultural and Processed Food Products Export Development Authority" (APEDA). Evidence from Indian Agricultural Research Institute, New Delhi, and Directorate of Rice Research, Hyderabad, proved that the said 16 claims of the US firm Rice Tec Co. were similar to traditional Indian Basmati rice. Rice Tec Co. had to subsequently withdraw these 16 claims and limit their patent claims to only those for genetically worked germplasm (involving inventive human intervention), and with such limited claims, which did not match the Basmati characteristics related to its GIs, the patent was issued to Rice Tec Co. on September 2,1997 (Basmati Patent US5663484, Rice Tec in).

1.5.4 PROTECTION OF PLANT VARIETIES AND FARMERS RIGHT ACT (PPV &FR ACT) 2001

The TRIPS Agreement further prescribed for the protection of new plant varieties as a form of IPR either by a patent or by an effective sui generis system or by any combination thereof. The Indian Patent Act 1970 excluded agriculture and horticultural methods of production from patentability, and hence, a sui generis system for the protection of plant varieties was legislated in the form of Protection of Plant Varieties & Farmers Right Act 2001 (*http://www.plantauthority.gov.in/*) (PPV & FR Act) integrating the rights of breeders, farmers, and village communities, encouraging the development and cultivation of new varieties of plants and taking care of the concerns for equitable sharing of benefits. The PPV & FR Act covers all categories of plants, except microorganisms (*www.plantauthority.gov.in_List_of_Cerificates.pdf*). Some of the major provisions of the PPV & FR Act are summarized as:

The IPR granted under the PPV & FR Act 2001 constituted a dual right–one is for the variety and the other is for the denomination assigned to it by the breeder. The rights granted under this Act are heritable and assignable, and only registration of a plant variety confers the right.

Varieties that can be registered under the said PPV & FR Act 2001 include the following:

1. **New Variety** that conforms to the criteria of novelty, distinctiveness, uniformity, and stability. A variety is considered to be
   (a) ***novel*** if, at the date of filing of the application for registration for protection, the propagating or harvested material of such variety has not been sold or otherwise disposed of by or with the consent of its breeder or his successor for the purposes of exploitation of such variety—(*i*) in India, earlier than 1 year or (*ii*) outside India, in the case of trees or vines earlier than 6 years, or in any other case, earlier than 4 years,
   (b) ***distinct***, if it is clearly distinguishable by at least one essential characteristic from any another variety whose existence is a matter of common knowledge in any country at the time of filing of the application.
2. **Extant Variety** means a variety that is
   (a) Notified under Section 5 of the Seeds Act 1966 (seednet.gov.in/PDFFILES/Seed_Act_1966.pdf) (54 of 1966): [Section 5 of Seeds Act prescribes ***Power to notify kinds or varieties of seeds***: *If the Central Government, after consultation with the Committee, is of opinion that it is necessary or expedient to regulate the quality of seed of any kind or variety to be sold for purposes of agriculture, it may, by notification in the Official Gazette, declare such kind or variety to be a notified kind or variety for the purposes of this Act and different kinds or varieties may be notified for different States or for different kinds or varieties may be notified for different States or for different areas thereof.*]

   (b) A farmers' variety.
   (c) A variety about which there is common knowledge.
   (d) Any other variety that is in the public domain.
3. **Farmer's Variety** means a variety that
   (a) has been traditionally cultivated and evolved by the farmers in their fields; or
   (a) is a wild relative or land race of a variety about which the farmers possess common knowledge.
4. **Essentially Derived Varieties (EDV)** in respect of a variety (the initial variety), shall be said to be essentially derived from such initial variety when it:
   (a) is predominantly derived from such initial variety, or from a variety that itself is predominantly derived from such initial variety, while retaining the expression of the essential characteristics that results from the genotype or combination of genotype of such initial variety;
   (b) is clearly distinguishable from such initial variety; and
   (c) conforms (except for the differences that result from the act of derivation) to such initial variety in the expression of the essential characteristics that result from the genotype or combination of genotype of such initial variety.

   EDV thus meant a variety that has been essentially derived from existing variety by any of the following means:
   (a) genetic engineering;
   (b) mutation;
   (c) tissue culture derived;
   (d) back cross derivative;
   (e) any other (ploidy change, etc.).

Under the said PPV & FR Act 2001, the period of protection for field crops is 15 years and for trees and vines is 18 years, and for notified varieties, it is 15 years from the date of notification under Section 5 of Seeds Act 1966. The rights granted under the Act are exclusive right to produce, sell, market, distribute, import, and export the variety (Table 1.1) (*www.plantauthority.gov.in_List_of_Cerificates.pdf*).

TABLE 1.1 Some Examples of New Plant Varieties Registered Under the Indian PPV & FR Act

Category of variety	Denomination of the candidate variety	Crop	Application no.	Date of certificate
Extant	Pratap Hybrid Maize 1	Maize	E20 ZM42 07 315	12/February/2009
Extant	Utkarsh (IPR 98-5)	Kidney Bean	3 PV3 07 202	12/February/2009
Farmer	Tilak Chandan	Rice	01 OS34 07 126	21/December/2009
Farmer	KUDRAT 9	Wheat	F1 TA4 09 219	04/April/2012
New	MIM 601	Maize	N02 ZM02 07 034	21/October/2011
New	JRO 204	Jute	115 CO5 08 423	18/January/2013
EDV	VICH-5 BG-II	Tetraploid Cotton	ED1 GH69 10 210	21/November/2012

1.6 PATENT RIGHTS AS IPR IN AGRICULTURE

Patents constitute the most challenging and widely discussed and debated form of IPR especially when considered in relation to basic needs of human survival, environmental issues attached to agriculture-based inventions, or advancements, which involve the use of natural biological resources as well as the differences in the level of technical capabilities and priorities of various nations and its countrymen.

Thus, the TRIPS Agreement is considered to be very crucial in the field of patent, because the standard established by the TRIPS Agreement laid down the basic framework for patentable subjects and scope of coverage and exclusive privileges that the member countries should guarantee which include:

(a) basic standard for patentability and a limited list of exceptions to patentable subject matter;
(b) no discrimination as to the field of technology, the place of invention, and whether the products are imported or made locally in terms of the availability and duration of the patent right;
(c) compulsory licenses;
(d) the term of protection for 20 years;

(e) availability of judicial review process for any decision to revoke or forfeit a patent;
(f) the burden of proof in deciding whether a product was obtained by a patented process; and
(f) conditions concerning the disclosure of the invention in a patent application.

In the TRIPS Agreement, there are three permissible exceptions to the basic rule on patentability as follows:

The first one is for inventions contrary to morality; this explicitly includes inventions dangerous to human, animal, or plant life or health or seriously prejudicial to the environment. The use of this exception is subject to the condition that the commercial exploitation of the invention must also be prevented and this prevention must be necessary for the protection of public order or morality (Article 27.2).

The second exception is that members may exclude from patentability diagnostic, therapeutic, and surgical methods for the treatment of humans or animals (Article 27.3(a)).

The third is that members may exclude plants and animals other than microorganisms and essentially biological processes for the production of plants or animals other than non-biological and microbiological processes. However, any country excluding plant varieties from patent protection must provide an effective *sui generis* system of protection (Article 27.3(b)).

Another important direction to subjects for patentability resided under TRIPS Article 27.3(b), which required countries to grant patent protection to microorganisms, non-biological and microbiological processes. The TRIPS Agreement had an additional important principle: IP protection should contribute to technical innovation and the transfer of technology. Both producers and users should benefit, and economic and social welfare should be enhanced.

In India, the substantial issue of protecting the IPRs as Patent and Design Rights had started since 1911 under the Indian Patents and Designs Act 1911 by bringing patent administration under the management of Controller of Patents for the first time.

After independence, the Government of India keeping in view that India has been an agriculture-based country had constituted a select

Committee under the Chairmanship of Justice (Dr.) Bakshi Tek Chand, a retired Judge of Lahore High Court, in 1949 to review the patent law in India in order to ensure that the patent system is conducive to the national interest. Based on the recommendations of the said Committee, the 1911 Act was amended in 1950 (Act XXXII of 1950) in relation to working of inventions and compulsory license/revocation. In 1952 (Act LXX of 1952), an amendment was made to provide compulsory license in relation to patents in respect of food and medicines, insecticide, germicide, or fungicide and a process for producing substance or any invention relating to surgical or curative devices. Finally, in 1967, an amended bill was introduced that was referred to a Joint Parliamentary Committee, and on the final recommendation of the committee, the Patents Act, 1970 was passed. This Act repealed and replaced the 1911 Act so far as the patents law was concerned. However, the 1911 Act continued to be applicable to designs until the Designs Act 2000 was separately enacted.

The Indian patent regime under the said 1970 Indian Patents Act (pre-TRIPS era) differed in many ways from that agreed upon under the TRIPS Agreement. This was principally because India among all developing countries had formulated a liberal patents regime since 1970 that promoted the country's interests and initially opposed the inclusion of patent laws in the negotiations developed under WTO. The then Patents Act of 1970 drastically restricted the rights of patent holders in fields linked to basic needs. This is due to the fact that the adoption of the Patents Act, 1970 was based on a lengthy legislative process and careful consideration of the socio-economic impacts of the patents in sensitive fields such as health and food.

The TRIPS Agreement had thus been widely debated in India because of some valid reasons based on the facts that provisions in TRIPS related to the country's patent laws and obviously expected to affect major areas of the country's well-being—health, agriculture, research, etc. It was thus after extensive rounds of debate that India finally agreed to become signatory to the Agreement on TRIPS of the WTO in 1995 along with other developing countries with the hope that TRIPS regime will result in free flow of trade, investment, and technical know-how among the member countries by removing barriers that exist in the form of differences in the standards of IP.

To have an indication of level of changes that the TRIPS Agreement required in then prevailing patent laws in India (pre-TRIPS regime under the Act of 1970), which were tuned to the then socio-economic conditions and India's status as an agricultural-based country, the following comparative could be a relevant indicator (Table 1.2):

Thus, in order to comply with the TRIPS provisions, Indian Patent Laws were revisited, and substantial amendments were introduced in transitory phases through stage-wise amendment to the Patents Act 1970, 1999, 2002, and 2005.

The Patents (Amendment) Act 1999 created a transitory pipeline provision to introduce privileges of exclusive marketing rights (EMR) and established mail box applications for patents of pharmaceutical and agro-chemical patentable products from January 1, 1995. The applicants could be allowed EMR to sell or distribute these products in India, subject to fulfillment of certain qualifying conditions. This was relevant only for a period of 3 years from entering of the TRIPS provisions in the year 1995, and its implications and relevance are no more applicable.

The Patents (Amendment) Act 2002 implemented the TRIPS-required 20-year patent term, introduced provision for reversal of the burden of proof for process patent infringement and modifications to compulsory licensing requirements. Notably, the 2002 amendments to the Act, for the first time extended the scope of patentable subject matter to accommodate biotechnological inventions including microbiological processes. The Act

TABLE 1.2 Highlighting Features (Differences) of Pre-TRIPS Indian Patents Act 1970 and TRIPS

Indian Patents Act of 1970 Pre- TRIPS Regime	TRIPs Obligations
Only process patents was allowed; NO product patents in food, medicines, chemicals.	Both Process and product patents to be allowed in almost all fields of technology .
Term of patents: 14 years; 5-7 in chemicals, drugs.	Uniform Term of patents :20 years for all fields.
Compulsory licensing and license of right provisions.	Limited compulsory licensing, no license of right is provided.

Source: http://www.ipindia.nic.in/history-of-indian-patent-system.htm.

clarified that "chemical processes," which were already considered patentable but given a patent term of only 5–7 years, included "biochemical, biotechnological, and microbiological processes."

This amendment undoubtedly advanced the scope of patentable subject and thus stimulated scientists for new inventions in the field of agrobiotechnology to secure IPR protection.

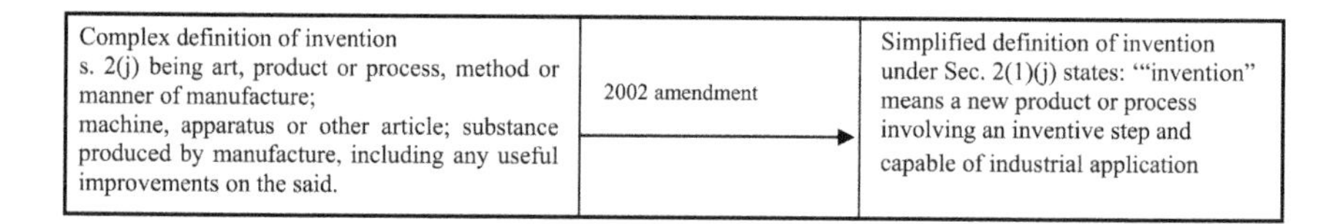

Complex definition of invention s. 2(j) being art, product or process, method or manner of manufacture; machine, apparatus or other article; substance produced by manufacture, including any useful improvements on the said.	2002 amendment →	Simplified definition of invention under Sec. 2(1)(j) states: '"invention" means a new product or process involving an inventive step and capable of industrial application

Importantly, the 2002 amendment of the Patents Act is also special as it is made mandatory to deposit the biological materials mentioned in the specification in the International Depository Authorities (IDA) under the Budapest Treaty 1977 on the International Recognition of the Deposit of Microorganisms for the Purposes of Patent Procedure.

The Budapest Treaty (WIPO, 1977) is an international convention governing the recognition of microbial deposits in officially approved culture collections, which was signed in Budapest in 1977 and later amended in 1980. The Treaty is aimed at overcoming the difficulties and on occasion of virtual impossibility of reproducing a microorganism from description in the patent specification and makes it essential to deposit the concerned strain in a culture collection center for testing and examination by others. It fulfills the need of describing a microorganism in the patent application and ensures obtaining samples of strains from the depository for further working on the patent.

The main feature of the Treaty is that a contracting State that allows or requires the deposit of microorganisms for the purposes of patent procedure must recognize, for such purposes, the deposit of a microorganism with any "international depositary authority," irrespective of whether such authority is in or outside the territory of the said State.

In practice, the term "microorganism" is interpreted in a broad sense, covering biological material, the deposit of which is necessary for the purposes of disclosure, in particular regarding inventions related to the food and pharmaceutical fields.

The Budapest Treaty was advantageous in that it eliminated the need to deposit the strains in each country in which patent protection is sought, because the Treaty provided for the deposit of a microorganism with any "international depositary authority," which suffices for the purposes of patent procedure before the national patent offices of all of the contracting States and before any regional patent office (if it recognized the effects of the Treaty).

The IDA is a scientific institution—typically a "culture collection"—which is capable of storing microorganisms. Such an institution acquires the status of IDA through the furnishing by the contracting State in the territory of which it is located of assurances to the Director General of WIPO to the effect that the said institution complies and will continue to comply with certain requirements of the Treaty. The deposits are made at an IDA in accordance with the rules of the Treaty on or before the filing date of the complete patent application. As of January 16, 2017, there are 46 IDAs in approximately 25 countries worldwide (WIPO, 1977).

Thus, the Budapest Treaty was directed to make the patent system of the contracting State more attractive because it is primarily advantageous to the depositor if he is an applicant for patent in several contracting States. The deposit of a microorganism under the procedures provided for in the Treaty was directed to simplify procedures and save money and also increase security because deposit is required only once, with one depositary authority. The Treaty increased the security of the depositor because it established a uniform system of deposit, recognition, and furnishing of samples of microorganisms (WIPO, 1977).

India acceded and ratified the Budapest Treaty on December 17, 2001. Currently, in India, Microbial Type Culture Collection and Gene Bank (MTCC) at the Institute of Microbial Technology (IMTECH), Chandigarh, and Microbial Culture Collection (MCC), Pune, are the two recognized international depositories of microorganisms.

1.7 NOTIFICATION BY THE INDIAN PATENT OFFICE (IPO) REGARDING DEPOSIT OF BIOLOGICAL MATERIAL

The Indian Patent Law duly incorporated the features of the Budapest Treaty with respect to biological material that is not available to the public, so that access to the material is available in the depository institution only after the date of the application of patent in India. The Controller General of Patents, Designs and Trademarks issued notification (http://www.ipindia.nic.in/writereaddata/Portal/News/159_1_115-public-notice) regarding the aforesaid, on July 2, 2014 which stated as follows:

According to the provisions of the Act, the deposition of such material in an International Depository Authority (IDA) under the Budapest Treaty shall not be later than the date of filing of patent application in India. However, the reference of deposition of biological material in the patent application shall be made within three months from the date of filing of such application as per Rule 13(8) of the Patents Rules, 2003.

The Patents (Amendment) Act 2005 (Act 15 of 2005) gave full effect to pharmaceutical/biotechnology product patents coverage mainly by amending Section 5 of the then Patent Act 1970 and making provision for the product patent in such fields of advancements in keeping with the TRIPS obligations.

The basic criteria for patentability of inventions in all fields in India as also reflected in the TRIPS Agreement included the basic triple test of qualifying subject for patentability as that having (a) "novelty," (b) "inventive step," and (c) "industrial utility."

The requisites of the above criteria for patentability usually followed globally can be appreciated from some related provisions of The Indian Patents Act 1970 as amended by the Patents Amendment Act 2005 currently followed is discussed in the following subsections.

1.7.1 NOVELTY (NOT ANTICIPATED)

- Section 2(1)(j) of the Indian Patents Act requires that the "invention" must be new, that is, it must be different from "prior art."
- Sec. 2(l) describes "new invention" as: any invention or technology which has not been anticipated by publication in any document or used in the country or elsewhere in the world before the date of filing of patent application with complete specification, that is, the subject matter has not fallen in public domain or that it does not form part of the state of the art.

- A prior art will be considered as anticipatory if all the features of the invention are present in the cited art.
- It further states that "the prior art should disclose the invention either in explicit or implicit manner."

1.7.2 INVENTIVE STEP (NON-OBVIOUSNESS)

- According to Sec. 2(1)(ja) of the Indian Patent Act, "inventive step" means a feature of an invention that involves technical advancement as compared to the existing knowledge or having economic significance or "both" and that makes the invention not obvious to a person skilled in the art.
- It is apparent from Sec. 2(1)(ja) that in order to fulfill the requirement of "inventive step," an invention should either have a technical advance or economic significance. Moreover, the invention should not be obvious to a person with ordinary skill in the art.

1.7.3 INDUSTRIAL APPLICABILITY

- Sec. 2(1)(j) of the Indian Patents Act 1970 prescribes that an invention to be patentable in India, should be capable of being industrially applicable.
- Sec. 2(ac) of the said Act further qualifies the meaning of "capable of industrial application" as, in relation to an invention, means that the invention is capable of being made or used in an industry.

Thus, while TRIPS Agreement laid down the basics of IP protectable subject matter, the member countries were free to prescribe the laws and provisions determining the scope and coverage of the same in its statutes and laws, and accordingly, the laws and rights and means of enforcement continue to be based on local laws in member countries.

Thus, in the USA, patents to plants are granted by the government to an inventor (or the inventor's heirs or assigns) who has invented or discovered and asexually reproduced a distinct and new variety of plant, other than a tuber propagated plant or a plant found in an uncultivated state provided it fulfills the basic requirement of patentability (35 U.S.C. 161) [https://www.uspto.gov/]. Plant varieties produced sexually (i.e., by seed) cannot be protected under the US patent law, but can be protected under a federal law enacted in 1970 known as the Plant Variety Protection Act.

In the area of plant breeding and genetic engineering, plants, parts, or products thereof are patentable under the European Patent Convention (EPC) as long as the feasibility of the invention is not restricted to a particular plant variety and if it fulfills the formal and substantive requirements of the EPC. Plant varieties on the other hand are specifically excluded from patentability by virtue of Art. 53(b) and Rule 27EPC, which provided that plants or animals are patentable if the "technical feasibility of the invention is not confined to a particular plant or animal variety" [Rule 27(b)) or "a product obtained by means of a technical process other than a plant or animal variety" (Rule 27(c))]. This is reflected from the recent decisions (dated March 25, 2015) of EPO's Enlarged Board of Appeal related to patentability of plants and plant products under the EPC rendered in G 2/12 (Tomato II) and G2/13 Broccoli II).

Two such exemplary illustrations of plant patents granted by the EPO include:

1. **"Tomato II"** (EP 1 211 926) a granted case entitled "Method for breeding tomatoes having reduced water content and product of the method" included product claims on "*A tomato fruit characterized by a capability of natural dehydration while on a tomato plant, natural dehydration being defined as wrinkling of skin of the tomato fruit when the fruit is allowed to remain on the plant after a normal ripe harvest stage, said natural dehydration being generally unaccompanied by microbial spoilage.*"
2. **"Broccoli II'** (EP 1 069819) the granted case involving a "Method for selective increase of the anticarcinogenic glucosinolates in Brassica species," claimed "*A broccoli inflorescence having elevated levels of 3-methylsulfinylpropyl glucosinolates, or 4-methylsulfinylbutyl glucosinolates, or both; wherein the broccoli inflorescence is obtained from a hybrid plant following crossing of broccoli breeding lines with wild species and, the levels of 4-methylsulfinyl-*

butyl glucosinolates, or 3-methylsulfinylpropyl glucosinolates, or both, are elevated above that initially found in broccoli breeding lines."

However, in contrast to such USS and EP procedures, in India, plant or plant part patenting is specifically prohibited under Section 3(j) of the Indian Patents Act, although protection of plant varieties can of course be protected under the sui generis system, that is, the Protection of Plant variety and Farmer's Right Act 2001 now in force in India. For example, a few extant varieties of maize produced by Indian Council of Agricultural Research (ICAR) registered under the PPVFR Act 2001 as Gujarat Makai-2 Reg. No.: E26 ZM49 07 336; Vivek Maize Hybrid-15 (FH-3176) Reg. No.: E50 ZM74 07 407; Pusa Extra Early Hybrid Makka-5 (AH-421) Reg. No.: E39 ZM61 07 386 (*www.plantauthority.gov.in_List_of_ Cerificates.pdf*).

Advancements in technology including agricultural sector (other than the new plant varieties discussed hereinbefore) qualifying as "inventions" under the law of patents also posed new challenges to the society and legal challenges of scope of its IPR protection. One of those challenges had been to identify those new technologies that may facilitate the mankind and environment without threatening the health and environment stability. Initially, agricultural patents throughout the world were restricted to inventions related to various agricultural machineries, which were followed by patents on agro-chemistry. Since the discovery of recombinant DNA technology in 1970s, biotechnology has emerged as an important research and industrial tool and the application of biotechnology to the field of agriculture has enabled inventions of diverse nature and good quality with industrial application. More specifically, the diverse nature of agro-biotechnological inventions really posed challenge to the existing IP laws and practice in the related technical domain. The major challenge to granting patent rights as a private property right on such advancements resided in the fact that biotechnology included the art of working/advancements related to the living organisms or their products (genes), which are usually regarded as occurring naturally and, therefore, faced with the question of patentability in involving such natural produces wholly or in parts thereof. Some of the relevant provisions of the currently followed Patents

Act in India, which lay down guidelines and extent of allowability of such advancements for the purposes of grant of patent in India, could exemplify the related challenges. According to Section 3 of the Indian Patents Act, the same is directed to limit the scope of subject matter eligibility for patentability of certain category of "inventions" that are also relevant for agricultural inventions (Indian Patent Act, 1970) (Table 1.3).

§3. *What are not inventions—the following are not inventions within the meaning of this Act—*

- *(a) an invention that is frivolous or that claims anything obviously contrary to well-established natural laws;*
- *(b) an invention the primary or intended use or commercial exploitation of which could be contrary public order or morality or that causes serious prejudice to human, animal or plant life or health or to the environment;*
- *(c) the mere discovery of a scientific principle or the formulation of an abstract theory or discovery of any living thing or non-living substance occurring in nature;*
- *(d) the mere discovery of a new form of a known substance which does not result in the enhancement of the known efficacy of that substance or the mere discovery of any new property or new use for a known substance or of the mere use of a known process, machine, or apparatus unless such known process results in a new product or employs at least one new reactant;*
- *(e) a substance obtained by a mere admixture resulting only in the aggregation of the properties of the components thereof or a process for producing such substance;*
- *(h) a method of agriculture or horticulture;*
- *(i) any process for the medicinal, surgical, curative, prophylactic diagnostic, therapeutic, or other treatment of human beings or any process for a similar treatment of animals to render them free of disease or to increase their economic value or that of their products;*
- *(j) plants and animals in whole or any part thereof other than microorganisms but including seeds, varieties, and species and essentially biological processes for the production or propagation of plants and animals;*

TABLE 1.3 Illustrations of Section 3 Prohibitions versus Any Possible Allowable Scope at a Glance

Section & Nature of invention	What is Non patentable	What May Be Patentable
3(b): Inventions causing serious prejudice to human, animal or plant life or health or to the environment.	Cloning method of animals; Process for modifying genetic identity of animals and animals resulting from such process; Process for preparing seed or other genetic materials comprising elements which might cause adverse environmental impact; Terminator gene technology which poses a serious threat to the animal and plant life as well as to the environment.	Pesticides, herbicides.
3(c): Mere discovery of a scientific principle or discovery of any living thing or non-living substances occurring in nature.	Microorganisms, nucleic acid sequence, proteins, enzymes, compounds (such as curcumin from turmeric) directly isolated from nature/traditionally known even if their function is unknown.	Processes of isolation and purification of these products and products from obtained microorganisms.
3(d): Mere discovery of a new form of a known substance which does not result in the enhancement of the known efficacy of that substance or the mere discovery of any new property or new use for a known substance or of the mere use of a known process.	Salts, esters, ethers, polymorphs, metabolites, pure form, particle size, isomers, mixtures of isomers, complexes, combinations and other derivatives of known substance.	Any such from/combinations with enhanced efficacy. Also, in case of such form or combination relates to drug/pharmaceutical substance then the significant and surprising efficacy should be related to "therapeutic" efficacy and no other efficacy data would then suffice.

TABLE 1.3 (Continued)

Section & Nature of invention	What is Non patentable	What May Be Patentable
3(e): A substance obtained by a mere admixture resulting only in the aggregation of the properties of the components thereof or a process for producing such substance	A composition containing components with their inherent property without any co-actions showing synergy.	A combination having the relationship between components as one of functional reciprocity or that they show a combinative effect (synergy) beyond the sum of their individual effects. Synergistic effect should be clearly brought out in the description by way of comparative trial run data illustrating the limitations of individual contributions of the respective actives and there summation vis-à-vis the surprisingly special and unexpected results achieved when the actives are combined together at the time of filing of the Application itself.
3(h): Method of agriculture or horticulture	Method of producing a plant, even if it involved a modification of the conditions under which natural phenomena would pursue their inevitable course (such as green house); Method of improving soil fertility by method of growing leguminous plants as inter-cropping; Method of producing mushrooms, Method for cultivation of algae. Methods of preparation of soil, sowing, applying manure and fertilizers, irrigation, protection from pests and weeds, harvesting and storage.	

Section & Nature of invention	What is Non patentable	What May Be Patentable
3(i): Any process for the medicinal, surgical, curative, prophylactic, diagnostic, therapeutic or other treatment of human beings or any process for a similar treatment of animals to render them free of disease or to increase their economic value or that of their products	Medicinal methods: Process of administering medicines orally, or through injectables, or topically or through a dermal patch. Surgical methods: Stitch-free incision for cataract removal. Curative methods: Method of cleaning plaque from teeth. Prophylactic methods: Method of vaccination; Therapeutic methods: Process of prevention and treatment of a disease. Increase economic value: Method of treating sheep for increasing wool yield.	Surgical, therapeutic or diagnostic instrument or device/apparatus/kit, vaccine. Treatment In plants.
3(j): Plants and animals in whole or any part thereof other than microorganisms but including seeds, varieties and species and essentially biological processes for production or propagation of plants and animals	Explanation: Naturally occurring microorganisms, Tissues, organs of plants and animals; Any process of manufacture or production relating to such living entities. Improved Breeds/ Strains of Animals/ Poultry/ Fish.	Genetically modified microorganisms and the process thereof; transgenic plant cells and process thereof; vaccines and the process of manufacture.
3(p): Traditional Knowledge or an aggregation or duplication of known properties of traditionally known components	Traditional knowledge available in public domain related to ayurveda, unani, siddha and yoga.	Compositions involving traditionally known components but showing unknown and unexpected synergy/therapeutic efficacy.

(p) *an invention that, in effect, is traditional knowledge or that is an aggregation or duplication of known properties of traditionally known component or components.*

The above prohibitory provisions under the Indian Patents Law and the possible scope of allowable subjects not attracted by such prohibitory clauses seem to balance between preservation and protection of natural produces and the nature and environment and to ensure that protectable subject matters are allowed exclusive privileges provided the same meets the qualification of patentability and do not in any way cause prejudice by way of its intended use or commercial exploitation being contrary to public order or morality or which causes serious prejudice to human, animal, or plant life or health or to the environment.

In the above backdrop of need for balancing patentable subject matter vis-à-vis natural bio resources and its inherent properties/characteristics, one of the most important and path-breaking patent cases that lead the way ahead for biotechnological inventions globally involving live entity, microorganism, is the legendary decision of US Supreme Court in *Diamond vs. Chakrabarty,* 477 U.S. 303 (1980), which was the first in the world to authorize microorganisms as a patentable subject.

In 1972, respondent Ananda Mohan Chakrabarty, a microbiologist, filed a patent application on the invention of *a bacterium from the genus Pseudomonas containing therein at least two stable energy-generating plasmids, with each of the said plasmids providing a separate hydrocarbon degradative pathway*. This human-made, genetically engineered bacterium, was capable of breaking down multiple components of crude oil. He developed this bacterium by engineering a way for multiple plasmids, each of which is able to break down different hydrocarbon components of the crude oil, to be incorporated into a single bacterium. The multiple plasmids would allow bacteria to break down oil from oil spills at a much quicker rate, and they would not be affected by environmental conditions. Because of this property, which is never possessed by any naturally occurring bacteria, Chakrabarty's invention was believed to have significant value for the treatment of oil spills.

Chakrabarty's patent claims were of three types: i) process claims for the method of producing the bacteria; ii) claims for an inoculum composed of a carrier material floating on water, such as straw, and the new bacteria; and iii) claims to the bacteria themselves.

The US patent examiner had initially allowed the claims falling into the first two categories, but rejected claims for the bacteria based on the argument: (1) that microorganisms are "products of nature" and (2) that, as living things, they are not patentable subject matter under 35 U.S.C. § 101 (MPEP Patent Laws, 2015): *Whoever invents or discovers any new and useful process, machine, manufacture, or composition of matter, or any new and useful improvement thereof, may obtain a patent therefore, subject to the conditions and requirements of this title.*

Chakrabarty appealed the rejection of these claims to the Patent Office Board of Appeals, and the Board had reversed in his favor, stating that "the fact that microorganisms are alive is without legal significance to the patent law." In response, Commissioner of Patents and Trademarks decided to take this case to the Supreme Court based on two basic arguments. The first relying on the Plant Patent Act 1930 and the Plant Variety Act 1970 to suggest that there is a congressional understanding about the terms "manufacture" and "composition of matter" under the Patents Act was not referring to living things. The second was that microorganisms cannot qualify as a patentable subject matter until Congress authorizes such protection since genetic technology was unforeseen when Title 35 U.S.C. 101 was first enacted.

On the other hand, the argument for Chakrabarty's claim to the bacteria relied on the fact that the invention was not to any natural phenomenon, but rather a manufacture or composition of matter, which is characterized as being a product of ingenuity that has "a distinctive name, character [and] use."

The Court interpreted the term "manufacture" as meaning, "the production of articles for use from raw or prepared materials by giving to these materials new forms, qualities, properties, or combinations, whether by hand-labor or by machinery," by the *Century Dictionary*. Additionally, "composition of matter" is accepted to include "all compositions of two or more substances and...all composite articles, whether they be the results of chemical union, or of mechanical mixture, or whether they can be gases, fluids, powders, or solids." Based on this, the patent law was accepted to be given a wide scope, which included Chakrabarty's microorganisms.

Subsequently, after a series of interpretations and discussions, the US Supreme Court finally concluded that the new bacteria were not "products of nature," because *Pseudomonas* bacteria containing two or more different energy-generating plasmids are not naturally occurring and issued the legendary decision to grant the patent (US4259444) in 1981, following which USPTO then extended the Supreme Court's decision to plants and non-human animals as well.

The legendary decision of US Supreme Court on *Diamond vs. Chakrabarty,* 477 U.S. 303 (1980), authorizing genetically modified microorganisms as a patentable subject had stimulated the IP system globally to consider a number of aspects that include the moral significance of treating such "inventions" involving plants, animals, microorganisms, and functional or structural components of life-forms including gene sequences, proteins, and cell cultures as "property."

In order to understand the kinetics of an invention leading to the development of biotechnology industry, it is worthwhile to mention Cohen-Boyer patent US4237224 granted in 1980 on recombinant DNA technology for the production of hepatitis B vaccine prepared from yeast (*Saccharomyces cerevisiae*). It is also important to state that this was the first ever and most important patent in biotechnology.

Stanley Cohen and Herbert Boyer combined their technical advanced findings for the said patent application:

- Boyer had isolated an enzyme that could cut DNA strings precisely into segments that carried the code for a predetermined protein and such segments could also be attached to other DNA strands.
- Cohen had developed a method for introducing antibiotic-carrying plasmids into certain bacteria, as well as method of isolating and cloning genes carried by the plasmids.

The main US Patent US423722 together with a patent for proteins produced using recombinant prokaryote DNA US4468464 and a patent US4740470 for proteins from recombinant eukaryote DNA defined the recombinant DNA technique of modern molecular biology. This technology gave birth to the biotechnology industry with launch of several biotechnology companies to practice such advancements (Feldman et al., 2007).

1.8 IMPACT OF THE DECISION OF THE U.S. SUPREME COURT DECISION IN INDIA

The US Supreme Court decision on *Diamond vs. Chakrabarty* also had its impact on the Indian Patent System, which was evident from one of the landmark decisions by the High Court at Calcutta in 2002 in respect of patenting of inventions involving microorganisms as live entities in an appeal case filed by Dimminaco A.G. against the decision of the Controller of Patents, India, refusing the grant to Dimminaco A.G. The historic judgment by the Calcutta High Court, in 2002 (Dimminaco vs. Controller of Patents and Designs & Others, 2001) in this case even prior to full-scale implementation of the TRIPS provisions under the Indian Patent Law pertained to the patentability of "living end-products of a biotechnological process" in the Indian system.

Dimminaco A.G., a Swiss company, had developed a live (attenuated) vaccine against Bursitis, an infectious poultry disease, and applied for patenting the process of its preparation. The application was turned down by the Indian Patent Office on the ground that the process did not constitute an "invention" under the Act based on the appreciation that the end product contained a living material and its procedure of development was only a natural process and hence not a patentable subject.

Based on the appeal by the Applicants, the Calcutta High Court had concluded to accept the process of manufacturing as patentable, even if the end-product contained a living organism.

The High Court at Calcutta explained that if the end product is a commercial and vendible entity, then it is patentable even if living organisms are present in it. The judgment followed the above discussed legendary judgment of US Supreme Court Diamond vs. Chakrabarty (1980), which granted patent to a genetically modified microorganism that could break down the multiple components of crude oil, a property not possessed by the naturally occurring bacteria. The High Court Calcutta order opened up new opportunities for obtaining patents in India on microorganism-related inventions involving agricultural biotechnology as well which were hitherto not granted. This milestone decision paved way for the patentability of numerous such inventions containing living microorganisms.

The post-TRIPS amended provisions of the Indian Patent Law currently in force categorically excludes non-natural occurring microorganisms from the prohibitory subjects of Section 3 subject to meeting other patentability requirements.

Apart from the challenges of patentability and the adaptability of the law to allow subjects such as microorganisms (genetically modified /not naturally existing), there has also been other related challenges globally on patentability of related agricultural subjects especially on the scientific development of new crop varieties, which began only in the last 50 years, following the discoveries and advancements in the field of genetics. Application of genetics into plant breeding resulted in major advances culminating in the development of high-yielding hybrids and varieties that have enhanced nutritional value in grains, which also fulfilled the concerns of the food security. Indian agriculture also started producing hybrids and high-yielding crop varieties in the 1960s, resulting in Green Revolution. However, although initially there was not much demand for ownership on plant varieties during those days, IP protection on such subjects had received enormous attention even in developing countries since the TRIPS discussions took effect. Modern biotechnological advances posed new challenges before the existing patent laws of many countries as biotechnological inventions in agriculture differed markedly from chemical and mechanical inventions that have been the traditional subject matter of patents for a long period.

One of the main features of modern agricultural biotechnology is its increasing proprietary nature. Unlike the agricultural inventions of the past, which came out of publicly funded labs, modern day biotechnologies are generated by private parties who have been advocating for patent protection in such advances as well being evolved by way of extensive research and developmental efforts along with substantial financial investments and such efforts unless recognized under law of patents by way of exclusivity in favor of the inventor/owner/proprietor was found to lead to the lack of incentive to work on such advancements at the cost of huge intellectual and financial efforts. Thus, the ownership of IPRs in agro-biotech inventions has over the years assumed importance in relation to the research activities in such fields of agricultural sciences. Since the early 1990s, most major research organizations globally were found actively strategizing the coverage of their advancements in various countries based on the scope and ambit of protection afforded in various jurisdictions that varied from country to country depending upon the patent laws codified in the respective regions.

Likewise, in various other jurisdictions, the patent laws in India also had evolved by gradual transformation in allowing subjects of patentability in the field of agricultural biotechnology, as would be clearly apparent from the following subjects of comparison between the earlier pre-TRIPS Patents Act 1970 regime followed in India and the current post-TRIPS Patents Act 1970 as amended by the Patents Amendment Act (2005) (Table 1.4).

Thus, despite the fact that TRIPS had brought a unified charter of basic minimum scope of coverage under IPR laws in member countries of the WTO, each country depending upon the socio-economic standings was free to evolve their IPR laws to meet its own national requirements and also to safeguard its own national interests and the public interest at large as well. Such development of local laws in keeping with the TRIPS obligations according to the national requirement and end objectives would

TABLE 1.4 Illustrations on Patentable Subject in India-Under the Pre-Trips and the Post-Trips Act:

Item	Pre-TRIPS Old Act (1970)	Post-TRIPS Act 1970 incorporating Amendment Act (2005)
Human Life/Genes	Not patentable	Not patentable
Animal varieties	Not patentable	Not patentable
Microorganisms	Not patentable	Patentable if Genetically Modified /not naturally occuring
Non biological processes	Patentable	Patentable
Non biological products	Not patentable	Patentable
Biochemical and biotechnological process	Patentable	Patentable
Bio products	Not patentable	Patentable
Crop/Plant varieties	No protection	Protection under PPV&FR Act
Method of Agriculture and Horticulture	Not patentable	Not patentable

be apparent from the fact that substantially the same invention originating from the same inventor/assignee/legal entity varied in scope of claiming allowable subject matter from country to country while all such countries may be members of the TRIPS/WTO. Therefore, it is extremely important to appreciate such variation in legal scope of possible coverage of variety of subject matter in any advancement in various jurisdictions/countries especially for effective coverage in agriculture-related advancements due to its inherent challenges as already discussed unlike for any other fields of advancement.

USA, being the pioneer in biotechnology research, does have great influence on the development of laws and related precedents in agricultural sciences, while the European Union reflects the unified approach of different member states in a politically diversified system. India represented the concerns of developing countries in general and the continuing need to balance patent grants against any form of public prejudice.

In the USA, any living organism that is the product of human intervention (such as by some breeding process or laboratory-based alteration) qualifies as a patentable matter under 35 U.S.C. 101 following the US Supreme court decision of *Diamond vs. Chakrabarty* (1980) 447 US 303. Furthermore, the USA extended patent protection to plants produced by either sexual or asexual reproduction and to plant parts including seeds and tissue cultures (*Ex parte Hibberd* (1985) 227 USPQ 433). US Patent and Trademark Office would consider non-naturally occurring, non-human multicellular living organisms, including animals, to be a patentable subject matter within the scope of 35 U.S.C. 101 (MPEP Patent Laws, 2015).

In Europe, individual plant varieties *per se* are not patentable; however, a plant that is characterized by a particular gene (as opposed to its whole genome) is not included in the definition of a plant variety and is therefore patentable. Thus, in Europe, transgenic plants are patentable if they are not restricted to a specific plant variety, but represent a broader plant grouping. The European Directive considers plant cells to be "microbiological products" and as a result are patentable under Article 53(b) EPC, which states "plant or animal varieties or essentially biological processes for the production of plants or animals; this provision shall not apply to microbiological processes or the products thereof."

Under the European Union, biological material isolated from its natural environment or produced by means of a technical process is considered to be the subject of invention even if it occurred in nature (EU Directive 1998/44EC.art 3.2). A process for the production of plants or animals which is based on the sexual crossing of whole genomes and on the subsequent selection of plants or animals is excluded from patentability as being essentially biological under Article 53(b) EPC. This applies even if the process comprises human intervention, including the provision of technical means, serving to enable or assist the performance of the process steps or if other technical steps relating to the preparation of the plant or animal or its further treatment are present in the claim before or after the crossing and selection steps. European patent also bars any invention against public order and morality under Article 53(a) EPC and bars invention on plants and animals along with naturally occurring processes.

1.9 THE INDIAN PATENT SCENARIO IN THE BIOTECH DOMAIN

Subsequent to the amendments made to the pre-TRIPS Patents Act, by way of amendments under the Indian Patent Act in 1999, 2002, and 2005, there has been a phenomenal growth in the number of patent applications, and consequently, an increase in the number of patents granted in the field of agriculture including agro-biotechnology. Despite the increase in the biotechnology patents, Indian legislature according to the currently followed provisions of the Patents Law does seem to be continuously very cautious in granting of patent on life forms.

The patenting activity in different sectors of agricultural science in India has shown a gradual increase since 2007 (post-full-fledged implementation of applicable TRIPS provisions under the patents law in India). There has been host of patents filed in areas of biocides, pest repellants or attractants, and plant growth regulators closely followed by applications in the area of new processes for obtaining plants cells and plant tissue culture; animal husbandry, silk rearing or breeding new animal breeds; and horticulture, cultivation, and forestry.

The trend of Indian patents in the various classes of agricultural science during the period of 2013–2016 reflects that out of around 3000 patents filed in different categories of agricultural technology, 16.22% were filed by Indian Applicants and 83.78% by the foreign Applicants (Source: Indian Patent office database). Maximum patent applications involved inventions on A01N (pest attractants, PGRs), followed by A01G (horticulture, forestry), A01D (harvesting, mowing), A01M (catching, trapping of animals), and A01B (soil).

1.10 NOTABLE PROVISIONS OF THE INDIAN ACT APPLICABLE TO AGRICULTURE-RELATED ADVANCEMENTS

In India, agro-biotech inventions like any other fields of advancement besides meeting the patentability requirements of novelty, inventive step, and industrial application must constitute patent-eligible subject matter so that it is not attracted by the provisions of non-patentable inventions specially prescribed under Section 3 of the Indian Patents Act. Some of the provisions of Section 3 prohibitions on patenting in India relevant for agriculture/biotech-based advancements include Sections 3(b), (c), (d), (e), (h), (i), (j), and (p) of the Indian Patents Act 1970 currently in force.

Section 3(b) prohibits advancement as not an "invention" if its use or commercial exploitation is "contrary to morality" or "causes serious prejudice to human, animal or plant life or health or to the environment." This prohibition can have relevance in relation to the agro-biotech advancements. Also, likewise alteration of genomic materials of an organism may have a far-reaching impact upon the human, animal, plant, and the environment. Hence, advancements involving process of cloning of animals, process of modifying genetic identity of animals that are likely to cause them suffering without any substantial medical or other benefit to mankind or animal, and process of preparing seeds or other genetic materials that might cause adverse effects are excluded from the patentability.

Section 3(c) disallows advancements as "invention" on "discovery of any living thing or nonliving substances occurring in nature." For example, the extracts and isolates of biological materials are generally considered as a mere discovery of a naturally occurring substance and is therefore barred under this provision. Thus, naturally occurring microorganisms, DNA, RNA, proteins, and any chemicals isolated from living organisms are considered not an "invention" under the Indian Act. However, advancements on the process of isolation of such substances can be patentable if it meets the three basic criteria of Novelty, Inventive Merit, and Industrial Applicability. The Indian Patent Office Guidelines on the Examination of Biotechnology Applications for Patents expressly state that sequences isolated directly from nature are not patentable. In fact, under existing jurisprudence, biological materials obtained as a result of substantial human intervention which distances from their natural counterparts are only considered patentable.

In this context, prosecution details of the Indian patent application 973/MUMNP/2010 may be relevant. Initially, the invention claimed as follows:

1. *An isolated promoter comprising an isolated polynucleotide having a sequence selected from the group consisting of Seq.ID No. 1 and a sequence that hybridizes to Seq. ID No. 1 under conditions of defined stringency.*
2. *The polynucleotide according to claim 1 wherein the conditions are low stringency.*
3. *The polynucleotide according to claim 1, wherein the said conditions are high stringency.*
4. *The promoter of claim 1 wherein the said promoter drives the transcription of an operably linked gene in a plant non-endosperm tissue.*

The above claims 1–4 were objected under the above discussed Section 3(c) by the Indian Patent Office. During prosecution the Applicant amended the claim 1 and deleted claims 2–4; the amended claim 1 reads as:

An isolated promoter consisting of an isolated polynucleotide having sequence selected from the group consisting of Seq. Id No 1 and a sequence that hybridizes to Seq. ID No. 1 under high stringency hybridization conditions, wherein said high stringency conditions are repeated washing for 30 minutes at 65^{0}C in a solution comprising 2XSSC (300 mM NaCl, 30 mM sodium citrate, pH 7.0) 0.5% (w/v) SDS solution.

It was further argued that the amended claim as now directed toward a promoter and expression cassette comprising polynucleotide sequence was not isolated from nature nor do they have characteristics as that of naturally occurring gene sequences.

However, Applicant's amendments and above argument in this context were not found to be persuasive by the Controller, because an isolated polynucleotide sequence included in the amended claim was isolated from plant genomic DNA, in particular, the nucleotide sequence was isolated from an *Oryza sativa* gene as described in the specification and thus what was being claimed in said claims well attracted the prohibitory scope of Section 3 (c) of the Act.

It may be relevant as well to traverse prosecutions of some cases to try to better appreciate some subjects of possible exclusion from the prohibitory scope of Section 3(c). On this, it may be relevant to mention about Indian application 787/MUMNP/2010 claiming a *recombinant vector comprising a nucleic acid encoding the protein Seq ID No 3 operably linked to a plant promoter, wherein the nucleic acid comprises of Seq No. 1 and 2,* which was allowed for grant (IN260860) by the Indian Patent Office.

The initial claims 1–3 were not allowable u/s 3(c) of the Patents Act, 1970 because these claims were directed to the subject-matter (DNA and amino acid sequence) occurring in the nature. Objected claims are as follows:

1. *A DNA sequence set forth in SEQ ID No. 1 for improving salt and drought tolerance to a plant.*
2. *A DNA sequence set forth in SEQ ID No. 2 for improving salt and drought tolerance to a plant.*
3. *An amino acid sequence encoded by the DNA sequence of claim 1, which is set forth in SEQ ID NO: 3.*

The amended claim directed to "a recombinant vector comprising the nucleic acid molecule encoding the protein of SEQ. ID NO 3" was later allowed since the claimed subject "a recombinant vector" was not existing naturally and was the product of recombinant DNA technology that involved human intervention, thus qualifying the same as a patentable "invention."

Section 3(d) of the Indian Patent Law currently followed is also frequently cited in case of modifications of existing substance as the said

Section of the Indian law prohibited "a new form of a known substance" from patentability if the claimed modification of an existing substance/a "new form of a known substance" failed to exhibit "enhancement of the known property/ efficacy." It is important to mention that this Section governing patentability of "new form of a known substance" needs special mention, because by way of a significant and enlightening decision of the Supreme Court of India it was clarified and held that the required "enhancement of the known property/efficacy" required a stricter evaluation in case of drugs and pharmaceutical product forms and claims based thereon and such efficacy in such drugs and pharmaceutical product forms should essentially clearly demonstrate significantly special "therapeutic" efficacy, and in the absence of surprising and significant therapeutic efficacy, the new form cannot qualify as an "invention" under Section 3(d). The relevant landmark case was traversed by the Supreme Court of India in the matter of an application for patent 1602/MAS/1998 in India made by Novartis on the new product form, the beta crystalline form of Imatinib Mesylate, a drug used to treat chronic myeloid leukemia (CML), a type of blood cancer marketed under the names "Glivec" or "Gleevec."

Supreme Court of India on April 1, 2013, delivered a landmark judgment rejecting Novartis's 1998 Indian patent application for said beta-crystalline form of Imatinib Mesylate, with the observation on special requirement of therapeutic efficacy to qualify as an "invention" under Section 3(d) and disregarding the data on bioavailability relied upon by the Applicants to justify efficacy of the new drug form. The relevant observation of the Apex Court (Supreme Court of India, Civil Appeal Nos. 2706–2716, **2013**) in India which has now translated as a statutory requirement under the patent laws in India read as hereunder:

What is "efficacy"? Efficacy means "the ability to produce a desired or intended result." Hence, the test of efficacy in the context of section 3(d) would be different, depending upon the result the product under consideration is desired or intended to produce. In other words, the test of efficacy would depend upon the function, utility or the purpose of the product under consideration. Therefore, in the case of a medicine that claims to cure a disease, the test of efficacy can only be "therapeutic efficacy.

"It may be noted that the text added to Section 3(d) by the 2005 amendment lays down the condition of "enhancement of the known efficacy."

Further, the explanation requires the derivative to "differ significantly in properties with regard to efficacy." What is evident, therefore, is that not all advantageous or beneficial properties are relevant, but only such properties that directly relate to efficacy, which in case of medicine, as seen above, is its therapeutic efficacy.

The above findings of the Supreme Court of India also ended Novartis 8-year battle through various Indian legal forums to get its new form of the drug beta-crystalline form of Imatinib Mesylate patented in India (Application No.1602/MAS/1998).

In the same application based on the decision of the Controller of Patents initially refusing the application under said Section 3(d) in the absence of any data on efficacy of the new form of beta crystalline form of Imatinib Mesylate, in comparison with the already known form of Imatinib Mesylate (non-crystalline), Novartis had even challenged the constitutional validity of Section 3(d) in the light of the TRIPS Agreement, which according to the claimants did not stipulate such additional special requirement to prove efficacy to justify patentability. However, the High Court at Madras before which the constitutional validity of Section 3(d) was challenged upheld the decision of the Controller of Patents and confirmed that there is no conflict in provision of Section 3(d) and the Indian constitution or the TRIPS provisions .The decision upholding the constitutional validity of Section 3(d) and also of the Supreme Court of India's path-breaking directions on required more stringent qualification on "therapeutic" efficacy in case of new form of substances for use as drug/pharmaceutical products is now a standard requirement before the Indian Patent Office and needs to be strictly adhered to qualify new forms of drug/pharmaceutical preparations as an "invention." Thus, applicants for patent in India on such subjects should be careful that the disclosure in the specification on the priority date of such claim does disclose such special "therapeutic" efficacy in case claims are directed to any new form of an old/existing drug/pharmaceutical substance, as otherwise, such applications and claims although allowable in other jurisdictions may fail to qualify under the currently followed patent laws in India.

Another very important prohibitory arm of Section 3(d) of the Indian Patents Act mentioned above related to categorizing "new uses of known substances" as not a patentable "invention." A very important and notable case that appears to bring in the importance of such a safeguard provision under the Indian Act can be appreciated by way of the scope and legal implications of a granted EP Patent under European patent no 0728048 claiming new use of hessian, a natural produce.

The European Patent No. 0728048 was granted in 1999 claiming the use of hessian cloth/sheet to cover waste/dumping grounds as an "invention" thereby holding exclusive rights in almost all European countries including UK, Germany, Sweden, France among others on such use of hessian. The impact of the grant and patent exclusivity was such that Indian Jute industry who were one of the major suppliers of hessian in the European region were threatened with infringement of the granted EP Patent as contributory infringers on hessian exported to the European region. Jute Manufacturers' Development Council, a statutory body under the Textiles Ministry and Indian Jute Industries Research Association in order to protect the interest of the Indian Jute Industry in consultations with their patent lawyers after careful evaluation and being convinced on the apparent lack of inventive subject in the claimed use of the hessian decided to file an application for revocation of the granted patent on the ground that the claim on "new use of Hessian" had been well anticipated and/or lacked inventive merit from well-known use of jute, a hessian variety in India for several decades before the alleged patent claim in Europe.

The Patentees on the other hand had applied for the right claiming that it had invented the use of hessian cloth or sheet to cover waste or dumping grounds. Patentees had sold around 5.6 million square meters of hessian between 1995 and 1996 to some European countries including the UK, Germany, France, and Sweden on exclusive marketing right basis. The Patentees had been charging royalty on the cost of hessian for using it for commercial purposes to cover dumps and fill wasteland in Europe.

It was after a heavily contested proceedings before the EPO's opposition division which initially held the patent lacking in both novelty and inventive merit and later under Appeal which was also contested, the Appeal Board held the use of hessian (a jute variety) for such purposes given the documented use of jute for like purposes lacked in patentable substance. Thus, the EPO's Technical Board of Appeals upheld the earlier decision of its opposition division leading to final revocation of the patent right.

The revocation of European patent no.0728048 and related patent right helped to revive India's hessian exports to Europe, which was badly hit due to patent exclusivity. This case is a definite landmark case that brings out the importance attached to patents and need for proper techno-legal evaluation of subjects of patent claims and the required conscious effort of the trade and industry to be watchful on nature of rights and how IP/ Patents can drive the market dominance in India and abroad based on the variance in possible scope and allow ability in various jurisdictions.

Yet further notable case on this prohibitory arm of Section 3(d) was traversed in the matter of an application for patent (2407/DELNP/2006) in India by Monsanto claiming "a method of producing a transgenic plant with increased heat tolerance, salt tolerance, or drug tolerance" which was rejected by the Indian Patent Office, as the cold tolerant property of a cold shock protein (CSP) involved was already known in the state of art. Monsanto had argued that the invention did not claim any "new" use of known substance (i.e., CSP); instead, it submitted that the invention relates to a "method" of producing a "new product" (i.e., stress-resistant plants). Further, it adduced post filing data that demonstrate "superiority" of transgenic plants produced using the claimed method vis-à-vis wild plants.

However, upon appeal from the Controller's order for refusal of grant, the Intellectual Property Appellate Board (IPAB), the appellate authority against decisions of Controller of Patents in India, rejected the arguments reiterating the decision of the Indian Controller that the cold tolerant properties of CSPs are already known. The Board observed that the application in essence claims "new use" of specific proteins from "cold shock domain" for producing desired traits and therefore ineligible under Section 3(d).

Indian patent application 5583/DELNP/2010 claimed *An agent for improving growth of rice seedlings comprising, as active ingredients, 5-aminolevulinic acid represented by the formula (I)* 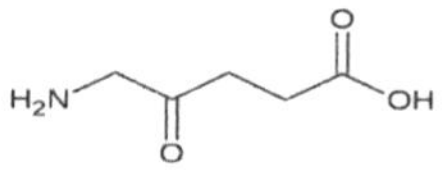 H_2N O O OH *a derivative thereof or a salt of the acid or the derivative, and a gibberellin biosynthesis inhibitor: ....wherein R1 and R2 each independently represent a hydrogen atom, an alkyl group, an acyl group, an alkoxycarbonyl group, an aryl group or an aralkyl group; and R3 represents a hydroxyl group, an alkoxy group, an acyloxy group, an alkoxy carbonyloxy group, an aryloxy group, an aralkyloxy group or an amino group.*

The claimed advancement was initially objected under Section 3(d) as being a new form of a known substance 5 aminolevulinic acid, which is the first compound in the porphyrin synthesis pathway (the pathway that leads to heme in mammals and chlorophyll in plants). However, it was clarified based on the examples from the specification disclosure that the claimed agent for improving growth of rice seedlings comprising, as active ingredients, 5-aminolevulinic acid represented by formula (1), a derivative thereof or a salt of the acid or the derivative, and a gibberellin biosynthesis inhibitor exhibited unexpected effect in the improved growth of rice seedlings in terms of increase in root weight, tillering number, increase in stem thickness, and greenness, which was beyond the summative effect of the individual components. The suitably amended claim incorporating the technical details was found to overcome the objection and the claims under the application on such lines was finally granted.

Amended and as finally granted claim under Indian Patent 267619 read as:

An agent for improving growth of rice seedlings comprising as active ingredients, 5 aminolevulinic acid or a derivative thereof represented by formula (1) or a salt of the acid or the derivative and a gibberellins biosynthesis inhibitor: $R^2R^1NCH_2COCH_2CH_2COR^3$ *wherein* R^1 *and* R^2 *represent a hydrogen atom and R3 represents hydroxyl group ......wherein the gibberellins biosynthesis inhibitor is at least selected from the group consisting .... and wherein the amount of the gibberellins biosynthesis inhibitor is 60.*

Section 3(e) excludes from patentability "a substance obtained by mere admixture" as not an "invention" under the Act.

Granted Indian Patent 271924 (*1627/DEL/2009)* entitled *Synergistic insecticidal and larvicidal botanical composition* claimed a novel insecticidal and larvicidal botanical synergistic composition comprising neem (*Azadirachta indica*) and *kabuli keekar* (*Prosopis juliflora*) plant pods extracts. In the First Office Action, the claims were objected under Section 3(e) of the Indian Patent Act on ground that the composition amounted to a mere admixture.

It was thereafter clarified to the Controller that Neem is known for insecticidal activities but kabuli keekar did not have any known insecticidal activity. Thus the finding that combination of *kabuli keekar* pod extract (KP) with neem kernel extract (NKP) in a specific ratio, was responsible for the surprising enhancement of the insecticidal and larvicidal activities of the neem as illustrated in the given examples 1–6 of the as filed specification and data under Table 1.5 demonstrated a surprising and unexpected synergistic effect and such data was relied upon to qualify the synergy in the claimed composition and avoid objection of mere admixture.

The above data in the specification demonstrated that the special effects of the selective combination of KP extract with NKP in a specific ratio and qualified the formulation as an "invention," thus overcoming the objections of 3(e).

Section 3(h) of the Indian Patent Act prohibited "a method of agriculture or horticulture" as an "invention" under the Act.

Indian application 9827/DELNP/2007 claiming a method of reducing mycotoxin contamination of a plant or harvested plant material that involved seed treatment with chemicals before/during sowing in the field for the cultivation of the said plant was not considered as an "invention" under Section 3(h) as being a process of agriculture.

The objected claim was read as:

A method of reducing mycotoxin contamination of a plant and /or harvested plant material said method comprising:

(a) treating plant propagation material with fludioxonil,
(b) germinating or growing said plant propagation material to produce a plant, and

TABLE 1.5 Synergistic Effect on Bioefficacy of KP and NKP on Mosquito Larva

Sl No.	Composition of the samples	% mortality of the larvae in 48 hrs
1.	NKP 1%	40%
2.	KP 1%	At par with control
3.	NKP+KP in the ratio of 50:50	70%
4.	NKP+KP in the ratio of 75:25	65%
5.	NKP+KP in the ratio of 25:75	25%
6.	Control	10%

(c) *harvesting plant material from said plant.*

The amended version of claim finally submitted by the Applicant was read as:

(a) *method of reducing mycotoxin contamination of a plant and/or harvested plant material obtained by germination or growing of a plant propagation material said method comprising a) treating plant propagation material with fludioxonil, and*

(b) *optionally treating the plant propagation material with insecticides, fungicides, bactericides, nematicides, molluscicides, bird repellents, growth regulators, biological agents, fertilizers, micronutrient donors or other preparations that influence plant growth such as inoculants or mixtures thereof.*

The steps for germinating or growing the said plant propagation material in (c) and harvesting plant material from the said plant in (d) of the original claims were deleted in the final amended claim. However, the invention was found to be still attracting Section 3(h) prohibition and was thus rejected which is reflected in the Controller's decision as:

The method claimed in the present invention is related to agricultural techniques which involve the treatment of seeds with the chemicals before sowing or during sowing in the field for the plant cultivation process.

As the present method involves the step of germination of the treated seeds to see the effects of the treatment on the harvested plant material thus the method clearly involve the agricultural techniques that are routinely used in the agriculture by the farmers for the plant protection by applying chemicals on the seed before sowing.

Thus, the subject matter involves agricultural technique and is not patentable subject matter.

Section 3(i) of the Indian Patents Act excludes from patentability *any process for the medicinal, surgical, curative, prophylactic, diagnostic, therapeutic, or other treatment of human beings or any process for a similar treatment of animals to render them free of disease or to increase their economic value or that of their products*. Significantly, the above provision therefore provided avenues for grant of patents for *in vitro* diagnostic methods performed on tissues or fluids that had been permanently removed from the living body, while the section strictly prohibited advancements on method of treatment and diagnosis of any disease among animals from qualifying as "invention."

However, such processes on plants are patentable, which is an important enhancement of scope of patentability under the Indian Patent Scenario in the post-TRIPS regime as reflecting from the amendment of the related Section under 3(i) by the amendment Act 38 of 2002. On the other hand, in compliance with TRIPS Article 27.3, which allows member states to exclude method of treatment claims, India categorically excluded method of treatment claims on animals and humans. It may be relevant to mention here for appreciating the variable scope of patentable subjects that some other patent office that prohibit method of treatment claims such as the European Patent Office allows some alternative forms of claims such as the "use of compound X in preparation of a medicament for treating disease Y" or "compound X for use in treating disease Y." In India, however, "use claims" are statutorily non-patentable claims, and claims need to be limited to new products or processes only. Accordingly, while under the EPC, you may have valid claims such as European Patent 1656150 claiming the "Use of cells of at least one bacterium strain of the *Bacilli* class for preparing a composition intended for preventing the vertebral compression syndrome in a fish belonging to the salmonid family," in India, such forms of claims are discouraged and disallowed.

The provision of the Patent Law in India currently in force which is possibly the biggest hurdle to the patentability of agricultural biotech inventions is Section 3(j), which excludes from patentability "plants and animals in whole or any part thereof," "seeds, varieties and species," and "essential biological processes for production or propagation of plants and animals." Accordingly, methods of crossing and breeding, which are essentially biological processes, are not patentable. However, parallel legislation—the Plant Varieties Protection and Farmers' Rights Act 2001—as already discussed above provides some scope of alternative mechanism of protection to transgenic plant varieties.

A decision dated July 8, 2013 of the IPAB in *Monsanto Technology LLC vs. Controller of Patents and Designs* is interesting in the context of Section 3(j). Monsanto Technology LLC applied for a patent method of producing a transgenic plant that was capable of withstanding harsh environmental conditions. Because the production of the transgenic variety involved substantial human intervention in inserting the rDNA molecule into the plant cell and transforming the cell into a climate-resistant plant, the invention was patentable according to the Applicant. However, the IPO was not convinced and held that the invention claimed related to an essentially biological process of regeneration and selection, which was excluded from patentability under Section 3(j) of the patent statute. The IPAB (2013) unequivocally clarified that the amended claimed method (Claim 1) relates to a method that requires several steps that together provided the claimed solution. It is a method that includes an act of human intervention on a plant cell and producing in that plant cell some change, and consequently fell outside the prohibitory scope of Section 3(j) although the grant was not allowed due to lack of inventive merit still attracting Section 3(d) as discussed earlier.

Objections under Section 3(p) of the currently followed Indian Patent Act categorically excludes from patentability as an "invention" which in effect "is traditional knowledge or which is an aggregation or duplication of known properties of a traditionally known component or components." To clear the qualifying bar set by this provision, claims are examined against searches of traditional knowledge databases, including the Traditional Knowledge Digital Library (TKDL), a digital data bank on traditional knowledge providing information on traditional knowledge in the country related to ayurveda, unani, siddha, and yoga developed by the Government of India and now shared with major Patent Offices all across the world. The database information is available in five languages (English, German, French, Japanese, and Spanish), which are used to evaluate the patentability of claimed advancements relating to traditional knowledge especially attached to natural bio-resources. Inventions that typically come under the scanner for ineligibility are extracts and alkaloids and active ingredients that are naturally present in plants, combinations of plants with known therapeutic effects, combination products of known active ingredients, and discoveries of optimum or workable ranges of traditionally known ingredients through routine experimentation.

Some case precedents relating to claims involving agricultural produce-based advancements explain the importance of TKDL content in evaluation of patent claims even in international domain.

- **The Turmeric Case**: In 1995, two non-resident Indians at the University of Mississippi Medical Centre, Jackson, were granted the US patent (patent number 54015041) for turmeric to be used for healing wounds (*http://www.uspto.gov*). The Indian Council for Scientific and Industrial Research (CSIR) filed a re-examination case with the US Patent Office challenging the patent on the grounds of "prior art," that is, existing public knowledge. The claim had to be backed by written documentation claiming traditional wisdom. The CSIR submitted a document proof in the form of research paper published in 1953 in the Journal of the Indian Medical Association revealing that the healing properties of turmeric had been in use for thousands of years in India.

  The US Patent Office upheld the objection and canceled the patent in 1997 as the turmeric advancement so far as claimed and patented that failed to meet the novelty and non-obviousness criteria based on documented prior traditional knowledge on inherent characteristics of turmeric.
- **The Neem Case**: Neem (*Azadirachta indica*) is a tree from India and other parts of South and Southeast Asia. Neem is used as natural medicine, fungicide, pesticide, and fertilizer and so, it attracted a considerable amount of international interest. In 1994, the EPO had granted a Patent (EP0436257) to the US chemical multinational company WR Grace & Co. and USDA for the preparation of a fungicide derived from the seeds of the Neem. Prior art search revealed that Indian farmers were using this knowledge for a long period, and several research studies had been conducted on the fungicidal property of Neem before the EP patent had been granted. EPO subsequently revoked the patent in 2005.
- **The Melon Case:** Monsanto Patent (EP1962578) claimed melons with a natural resistance to plant viruses, derived from breeding without genetic engineering. The EPO had to revoke the patent held by Monsanto on melons (EP1962578) directed to "CLOSTEROVIRUS-RESISTANT MELON PLANTS" for technical reasons as it was reported under opposition proceedings that such resistance was detected naturally in Indian melons. Thus, the advancement was

held to be based on essentially biological processes for breeding and claimed plant varieties and revoked.

Valuable state of art reference as rich resource of TKDL helpful also to major patent offices throughout the world in evaluating patentable novelty is well apparent from the traversal of the case of Avesthagen Ltd., a Bangalore-based company which claimed under European Patent, bearing No. EP2152284 on the June 29, 2007, for a "synergistic ayurvedic/functional food bioactive composition." After an adverse search report from the EPO, based on TKDL references [D6: Mega Noikku Legyam TKDL, GP02/658, 1800, XP002615047; D7: Meganoikku Kudineer TKDL, GP02/666, 1800, XP002615048; D8: Neerizhiuvu Noikku Kudineer TKDL, AM05/1644, 1800, XP002615049], the present set of claims were not considered novel over these references by EPO. Avesthagen Ltd. did not reply to the EPO, therefore resulting in a deemed abandonment on January 6, 2012.

The interesting fact was that the above EP application claimed its priority from an Indian patent—1076/CHE/2007. The Indian application had gone through one round of examination, and the patent examiner raised several objections including prior art but could not make any reference to the TKDL because he then had no access to the TKDL. The patent agents for the patentee, instead of abandoning the application based on EP decision on January 24, 2012, that is, about 18 days after abandoning their EP patent, filed a reply to the Examination Report of the examiner defending the claims of the original patent application and seeking a grant of the patent. As the TKDL database was then not accessible by IPO, the same resulted in grant of the Indian patent.

However, on having access to TKDL and having knowledge of the fate of the corresponding application in EPO, the Central Government's vide a special notification had to revoke the erroneous grant in India under Section 66 on ground that the grant had been "prejudicial to the public." The relevant provision of Section 66 which safeguarded from subsistence of wrongfully granted patents in India read as under:

"Revocation of patent in public interest, where the Central Government is of opinion that a patent or the mode in which it is exercised is mischievous to the State or generally prejudicial to the public, it may, after

giving the patentee an opportunity to be heard, make a declaration to that effect in the Official Gazette and thereupon the patent shall be deemed to be revoked."

The above case therefore highlights the importance of TKDL and its relevance in relation to the prohibitory provisions under Section 3(p) of the Indian Patents Act.

Apart from meeting, the patentability requirements are to qualify "absolute novelty," that is, the subject of the claims are globally novel and not prior published anywhere in the world, and are "inventive," that is, non-obvious and involves an inventive step as also "useful," that is, industrially applicable as well as the above-discussed prohibitory clauses of Section 3 of the Indian Patents Act, the "quid pro quo" requirement in grant of patent by the state/any country is extremely relevant and also equally applicable to patents in the agricultural domain. The "quid pro quo" or the very purpose of grant of patent by any state/country has been to encourage (1) working of invention within the country so as to result in the establishment in the country of viable industrial environment that could facilitate employment generation and growth and capital of the country and (2) the detailed disclosure of the technical advancement under the patent specification especially its manner of working such that upon expiry of the protected term of the patent rights and exclusivity vesting with the Patentee, the public are enabled to work the invention themselves and in competition with each other.

With regard to the issue of required working of patents in India, the Indian Patent Law has always prescribed in favor of the requirement for working of granted patents in India for national development. However, the TRIPS Agreement prescribed that member countries may not burden the Patentee to work the invention locally and importation of the invented product to meet local demands should be considered equal to working of inventions locally under a granted patent in member countries.

The currently followed Indian Patent Law continues to give priority for working of patented inventions in India and the relevant provision under Section 83 mentions that patents are granted in India to encourage inventions and to secure that the inventions are worked in India on a commercial scale and to the fullest extent that is reasonably practicable without undue delay. Thus, such provisions of the local law in India for granted patents and its working in India continues to be important as the Indian law under Section 84 further provides that non-working of patented inventions in India to meet reasonable requirements even after the expiry of 3 years from the date of grant can be a ground for issuing compulsory licensing of the patent in favor of parties other than Patentee interested in working the patented invention in India. Accordingly, there is clear direction under the Indian law as on date for encouraging grant of patents in India for working of the patented inventions in India by the Patentee or on behalf of the Patentee which in turn is directed to employment generation and advancement of art in the related domain in India which equally applies to agricultural-based advancements as well.

This is followed by the other issue of required invention workable disclosure of the advancement in the patent specification in sufficient detail by way of disclosure of the best embodiment of the advancement to enable a person in the art to carry out the invention without undue trials and experimentations. This requirement also featuring as a "quid pro quo" requirement under the Indian law which is clearly prescribed under Section 10(4) of the Indian Act, which also applies to agricultural advancement and reads as hereunder:

1. Every specification, whether provisional of complete, shall describe the invention and shall begin with a title sufficiently indicating the subject-matter to which the invention relates.
2. Subject to any rules that may be made in this behalf under this Act, drawings may, and shall, if the Controller so requires, be supplied for the purposes of any specification, whether complete or provisional; and any drawings so supplied shall, unless the Controller otherwise directs be deemed to form part of the specification, and references in this Act to a specification shall be construed accordingly.
3. If, in any particular case, the Controller considers that an application should be further supplemented by a model or sample of anything illustrating the invention or alleged to constitute an invention, such model or sample as he may require shall be furnished before the application is found in order for grant of a

patent, but such model or sample shall not be deemed to form part of the specification.

4. Every complete specification shall
   (a) fully and particularly describe the invention and its operation or use and the method by which it is to be performed;
   (b) disclose the best method of performing the invention which is known to the applicant and for which he is entitled to claim protection; and
   (c) end with a claim or claims defining the scope of the invention for which protection is claimed;
   (d) be accompanied by an abstract to provide technical information on the invention:

Apart from the above requirement of disclosure of the best embodiment of the claimed advancement as a requisite disclosure in the patent specification for a valid grant, also of importance in relation to advancements involving biological material has been the additional obligation on Applicants for patents in India involving such subjects as prescribed under Section 10(4)(d)(ii), which states that if the applicant mentions a biological material in the specification which may not be described in such a way as to satisfy full disclosure and best embodiment disclosure requirement if such material is not available to the public, the application need to be completed by depositing the material to an international depository authority under the Budapest Treaty such as to fulfill the following obligations:

10(4)(d)(ii) *if the applicant mentions a biological material in the specification which may not be described in such a way as to satisfy clauses (a) and (b), and if such material is not available to the public, the application shall be completed by depositing the material to an international depository authority under the Budapest Treaty and by fulfill in the following conditions, namely:*

(a) the deposit of the material shall be made not later than the date of filing the patent application in India and a reference thereof shall be made in the specification within the prescribed period;
(b) all the available characteristics of the material required for it to be correctly identified or indicated are included in the specification

including the name, address of the depository institution and the date and number of the deposit of the material at the institution;

(c) access to the material is available in the depository institution only after the date of the application of patent in India or if a priority is claimed after the date of the priority;

(d) disclose the source and geographical origin of the biological material in the specification, when used in an invention.

Additionally, when any patent specification uses any biological material, it is important for Indian applicants to submit a declaration in the patent application form to the effect that *the invention as disclosed in the specification uses biological material from India and the necessary permission from the competent authority shall be submitted before the grant of patent*, which then becomes mandatory.

The competent authority whose permission is then necessary for use of any biological material from India in any technical advancement disclosed in a patent specification is the authority mentioned under the Indian Biological Diversity Act 2002, which aims to protect sovereign rights over genetic resources and specifically requires prior approval from the independent National Biodiversity Authority for applying for a patent for inventions that use biological material from India. Thus, according to Section 6(1) of the Biological Diversity Act, 2002 (BD Act), all applicants for IPRs including patents applied in India based on the research and information of biological resources sourced from India should obtain approval from the National Biodiversity Authority (NBA) before grant. In the instance, however, biological resource used in any invention is sourced from outside India, then the permission of the biodiversity authority is not applicable. Thus, for advancements made outside India and involving biological material outside India, while applying for a patent in India on the advancement, there is no need for obtaining any permission from the Indian Biodiversity Authority. The Patent Act along with the BD Act in India thus incorporate safeguard provisions to prevent misuse and unauthorized exploitation of biological resources by requiring approval from NBA for the access to the biological resources of India and benefit sharing from the utilization of the same.

Thus, in effect, the Indian Patents Law provisions and precedents based thereon as currently applicable in the post-TRIPS regime appears to be balanced and yet meeting international standards as on the one hand it does adhere to the minimum basic protection for patentable subjects prescribed under the TRIPS, while on the other hand it reflects the safe-guard provisions to ensure streamlined use of biological material relevant for advancements in agricultural products and processes. In fact, the amendment in Patent Laws in India over the years to accommodate TRIPS obligations especially in relation to bioresource-based advancements/agricultural inventions have provided for also extension of scope for coverage of patentable subjects in some respects as compared to the pre-TRIPS regime. A relevant case that highlights such transformation of allowable subject under the Indian law is reflected from the prosecution details of Indian application No. 1372/DELNP/2006. This application originally claimed "a method for producing a rose characterized by artificially suppressing the rose endogenous metabolic pathway and expressing the pansy gene coding for flavonoid 3',5'-hydroxylase having the nucleotide sequence shown in SEQ ID No. 1and 3 wherein the rose flavonoid synthesis pathway is suppressed by ..." Such claim was initially objected by the Indian Patent Office under Section 3(j) which disallowed plants and animals in whole or any part thereof other than microorganisms but included seeds, varieties, and species and essentially biological processes for the propagation of plants and animals.

It was, however, argued by the Applicant that the claim clearly included step of "artificially suppressing the rose flavonoid synthesis pathway and expressing..." which cannot be identified as "essentially biological process." In order to further distance the advancement from Section 3(j), the Applicants limited the scope of their claims by way of amended claims directed to a method of producing a transformed rose plant cell instead of the "plant" to overcome the objection under the section 3(j).

Finally, the claims which were found allowable when reading as:

A method for producing transformed rose plant cell, said method is characterized by artificially suppressing the rose flavonoid synthesis pathway and expressing pansy gene coding for flavonoid . . . having the nucleotide sequence shown in SEQ.ID No. 1or 3, wherein the rose flavonoid synthesis pathway is suppressed by (i) artificially suppressing expression of rose endogenous dihyroflavonol reductase, . . .

With regard to living organisms and patentability, as noted above, some provisions expressly exclude the patentability of sequences isolated directly from nature, but an exception may be carved out for biological materials that are obtained as a result of substantial human intervention, for example, the recombinant DNA cDNA. This is well explained in the landmark decision, the 2013 US Supreme Court decision in *Association of Molecular Pathology vs. Myriad Genetics, Inc* (569 U.S., 2013) saw the court unanimously hold that full-length isolated, naturally occurring DNA molecules/genes or gene fragments are not patentable, as these were naturally obtained "products of nature." However, cDNA–DNA molecules in which the naturally occurring non-coding regions (introns) are absent—were found to be patent eligible. The court reasoned that cDNAs do not occur naturally and are synthesized from RNA in the laboratory with sufficient human interventions, thus validating the patent eligibility of engineered/recombinant DNAs.

Importantly, it is also relevant to mention that under the pre-TRIPS regime of the Indian Patent Act, the provision under Section 3(i) read as "any process for the medicinal, surgical, curative, prophylactic or any other treatment of human beings or any process for a similar treatment of animals or plants to render them free of disease or to increase their economic value or that of their products." However, post-TRIPS, the same section as amended Section 3(i) currently reads as "any process for the medicinal, surgical, curative, prophylactic, diagnostic, therapeutic, or any other treatment of human beings or any process for a similar treatment of animals to render them free of disease or to increase their economic value or that of their products."

It would thus be clear from the above that had this application been examined under the pre-TRIPS provisions of Section 3(i) the claims directed to a method of producing transformed rose plant cell . . . as allowed under the current law in India would have faced further objection and statutorily prohibited. This therefore reveals the expansion of scope for patentability of related subjects under the presently applicable post-TRIPS provisions of the Indian law.

In yet another Indian patent Application No. 1425/DELNP/2004 on an advancement titled "Method for Modifying Plant Morphology, Biochemistry and Physiology Comprising Expression of Plant Cytokinin Oxidase" the same had initially claimed:

A method of increasing seed size or weight, embryo size or weight, cotyledon size which method comprises introduction and expression of an isolated nucleic acid encoding a cytokinin oxidase thereby increasing the level or activity of a cytokinin oxidase in a plant or plant part, preferably seeds, embryos, cotyledons; which comprises introduction and expression of an isolated nucleic acid selected from the group consisting of: . . . nucleic acids . . . . It was objected by the Indian Patent Office under the Section 3(j) being essentially biological process.

The applicant amended the claim to "*A method of introduction and expression of an isolated nucleic acid encoding a cytokinin oxidase thereby increasing the level or activity of a cytokinin oxidase in a plant or plant part preferably seeds, embryos, cotyledons and increasing size and weight thereof; said method comprising introduction and expression of an isolated nucleic acid selected from the group consisting of....*"

It was then clarified to the controller that the claimed invention is related to introduction and expression of an isolated nucleic acid encoding cytokinin oxidase in plant part, increasing size, and weight thereof and subsequent claims are related to promoter and controlled expression of cytokinin oxidase in particular plant part which do not come under Section 3(j) as essentially biological process for production or propagation of plants and animals. Proper interpretation of the legal provisions in place and qualifying the advancement by proper claim construction thus helped in overcoming the objection with subsequent grant of the patent as IN256867.

In another case under Patent Application 2212/MUMNP/2009 entitled "New Hybrid System for *Brassica napus*," the original claims were directed to: "*A method for producing or multiplying seed of a conditionally male sterile Brassica napus line with the genotype MsMSrfrf said comprising the steps of: i) providing a conditionally male sterile Brassica napus line with the genotype MsMSrfrf . . . ii) exposing said conditionally male sterile Brassica napus line for at least 4 hours to a temperature higher than 35°C iii) exposing the heat treated . . . to a temperature less than 33°C until development of male flowers iv) allowing for self-pollination of the B. napus plant . . .*

As would be apparent from the above principal method claim, the same was clearly directed to a process involving the production of/propagation of plant which included seeds. According to Section 3(j) of the Indian Patent Act, the above process was construed as essentially biological process for production or propagation of plants including seeds and hence not an allowable subject for invention.

The same application also included claim directed to part of a Brassica plant, which again clearly attracted Section 3(j) of the Indian Act as discussed above.

The above discussion would go to reveal the importance of appreciating the scope of allowable subject matter for patenting based on the local laws in TRIPS-member countries, which becomes extremely relevant because while the TRIPS Agreement basically prescribed the minimum standard of protection of IPR rights in various member countries, the flexibility to allow broader coverage and the extent thereof after complying to the minimum standards prescribed by TRIPS continued to vest in each member country and related local legislations.

A case in point which could further bring out such relevance of patentable subject matter and its allowability varying from country to country could be appreciated from the contents and scope of the invention claimed under Indian patent application no.: 2407/DELNP/2006[34] which claimed for the following:

Claim 1: "A method of producing a transgenic plant comprising the steps of: (a) inserting into the genome of plant cells a recombinant DNA molecule comprising a DNA encoding a CSP, wherein said DNA encoding said CSP is operably linked to a promoter and operably linked to a 3' transcription termination DNA polynucleotide; (b) obtaining transformed plant cell containing said recombinant DNA; (c) regenerating plants from said plant cells; and (d) selecting a plant for increased heat tolerance, salt tolerance, or drought tolerance."

The Indian Patent Office that examines applications in keeping with the Indian Patents Act in force had observed that such method for enhancing drought tolerance in plants and compositions thereof were objectionable based on statutorily and not patentable "invention" category prescribed under Section 3(j) [essential biological process] and Section 3(d) [new use of a known substance] as the CSP used in the method for enhancing the drought tolerance was well known in the art as was documented and, accordingly, the method and the composition involving such known CSP attracted the rejection as a mere new use of a known compound and accordingly cannot qualify as an "invention."

Upon refusal of the above application, an appeal was preferred before the IPAB, and the Appeal Board after traversing the relevant issues had observed that the method required several steps which involved significant human intervention on a plant cell and producing some genetic changes in the said cell. Therefore, according to the Appeal Board, the said method cannot be considered as the essentially biological process and did not attract Section 3(j). In spite of the exclusion of the claimed invention from the scope of Section 3(j), the Appeal Board had observed that the invention would not be considered as a patentable subject matter because the alleged invention related to the discovery of some additional application of the known CSPs in plants under section 3(d). Thus, it was held by the Appeal Board that such discovery is merely a discovery of a new use of known substances and not an invention.

In the above backdrop of prosecution in India under the Indian application 2407/DELNP/2006 leading to rejection in India, a perusal of the corresponding prosecution in USA and EPO reveal the allowability of the invention for patenting as statutory allowable subjects in other regions based on the local laws as apparent from the Table 1.6.

Another case that illustrates the importance of not only the subject matter of the claims and its varied scope of patentability in various regions/countries as apparent from the preceding case studies but even the importance of language of claim construction in converting a statutorily non-patentable subject into a possible statutorily allowable subject can be traversed in relation to Indian national phase application no.: 5699/DELNP/2007 based on PCT international application PCT/US2005/046013 (WO 2006069017) which was directed to the following claim:

"A plant cell with stably integrated, recombinant DNA comprising a promoter that is functional in plant cells and that is operable linked to DNA that encodes a protein having domains of amino acids in a sequence that exceed the Pfam gathering cutoff for amino acid sequence alignment with a Pfam Homeobox protein domain family and a Pfam HALZ protein domain family; wherein the Pfam gathering cut off for the Homeobox

TABLE 1.6 Comparison of Patent Application on Drought Tolerance Transgenic Plant Encoding CSP Submitted at USPTO, EPO, and Indian Patent Office

US7786353	EP 2371841	INDIA 2407-DELNP-2006
Granted	Granted	Refused due to new use of a known substance (Section 3 (d) prohibition).
1. Principal claim: A drought tolerant transgenic plant selected from the group consisting of corn, and rice that has been transformed with a DNA molecule that expresses a protein having the amino acid sequence of SEQ ID NO:63 or a protein having the amino acid sequence of SEQ ID NO:65.	1. Principal claim: A plant and the seed there from having inserted into the genome of its cells a recombinant DNA encoding a CSP, wherein said CSP comprises the cold shock domain sequence [FY]-G-F-I-x(6,7)-[DER]-[LIVM]-F-x-H-x-[STKR]-x-[LIVMFY] of SEQ ID NO:3, … wherein expression of said protein in said plant or a plant grown from said seed refers drought tolerance .....”	Not patentable
2. A transgenic propagule of the plant of claim 1.	2. The plant and the seed .... wherein said CSP is at last 80% identical across the length of seq ID. No. 1.	
3. A transgenic progeny of the plant of claim 1.	3. The plant and seed … where the CSP I a plant protein.	
4. The plant of claim 1 that has an increased yield when compared to a non-transformed plant of the same species and when said transgenic plant and said non-transformed plant are grown under drought stress.	4. The plant seed acc. To .... is a corn plant seed or a soybean plant seed or a cotton plant seed.	
5. A field crop comprising transgenic plants germinated from a propagule comprising a protein having the amino acid sequence of SEQ ID NO:63 or a protein having the amino acid sequence of SEQ ID NO:65, wherein said field crop is selected from the group consisting of corn and rice.	5. A plant cell having recombinant DNA inserted into its genome....	

protein domain family is -4 and the and the Pfam gathering cut off for the HALZ protein domain family is 17; wherein said plant cell is selected from a population of plant cells with said recombinant DNA by screening plants that are regenerated from plant cells in said population and that express said protein for an enhanced trait as compared to control plants that do not have said recombinant DNA; and wherein said enhanced trait is selected from group of enhanced traits consisting of enhanced water use efficiency, enhanced cold tolerance, increased yield, enhanced nitrogen use efficiency, enhanced seed protein and enhanced seed oil."

During examination in India, the claim was objected under Section 3(j) [essential biological process] and Section 3(h) [method of agriculture and horticulture].

However, it was possible to cover the same subject matter of the advancement in India by rewording the claims of the original PCT application on the following lines:

"A method of manufacturing non-natural transgenic seed that can be used to produce a crop of transgenic plants with an enhanced trait resulting from stably integrated recombinant DNA comprising .... the method of manufacturing said seed comprising: (a) screening a population of plant...... (b) selecting from said population one or more plants..... (c) verifying the said recombinant DNA..... (d) analyzing tissue of........ (e) collecting seed from the selected plant."

Controller's decision traversing waiver of the objection raised U/S 3(j) on the revised claims 1–6 observed that the said claims relate to a method of manufacturing non-natural, transgenic seed used to produce the plants demonstrating the traits such as increased yield, enhanced water use efficiency, enhanced nitrogen use efficiency, increased kernel weight, increased kernel size, or increased ear size. Said claims require the presence of a transgenic seed with a stably integrated, recombinant DNA molecule, which can be achieved only upon substantial human intervention and is not an essentially biological process.

On the issue of objection raised U/S 3(h) on the revised claims 1–6, the same was waived for the reason that the said claims related to a method of manufacturing non-natural, transgenic seed used to produce the plants demonstrating the traits such as increased yield, enhanced water use efficiency, enhanced nitrogen use efficiency, increased kernel weight, increased kernel size, or increased ear size. Said claims require analyzing tissue of a selected plant and the step of screening a population of plants for the enhanced trait, which involves a lot of empirical testing through specially designed methods. The said steps are not carried out by farmers or in conventional agriculture. In view of the above said, the application was granted under IN 272763.

Thus, the above would clearly reveal how claim construction plays critical role to make it conform to the statutory requirements of the applicable laws and to ensure that the language of the claims do not continue to attract non-patentable subject matter.

It is also important to mention that unlike many other jurisdictions under the Indian Patent Law, after filing of an application for patent, the scope for amending the disclosure and the claims are highly limitative in nature which applies equally to agriculture-based inventions. More specifically, Section 59 of the Indian Patent Act specifies that post-filing amendment should be necessarily limited to the amendment by way of disclaimer, correction, or explanation only and no amendment of the claims are allowed which amount to expansion of scope of the claims as originally filed. It is, therefore, extremely important to be aware of such special and limited scope of post-filing amendment of Indian Patent application under the applicable local laws in India and ensure that the original disclosure and claims do not pose a hurdle to effectively claim any advancement discussed in the specification and allowable under the laws in India and thus special care should be taken to avoid rejection of application for amendments at a later date.

The above would go to reveal the importance of technical advancements and its strategized and judicious coverage under various applicable IPR laws in various jurisdictions is extremely relevant to the agricultural product and method-based advancements as well.

1.11 THE FUTURE OF IPR IN THE FIELD OF AGRICULTURE

It is therefore well established that the advancements in agricultural biotechnology and related IPR has serious implications for law and society. Earlier IP laws, which were framed in industrial age, have proved to be inefficient in the present information age as because the trend of inventions has changed greatly. Present day advancements involve applications of technology which thus confronts existing patent laws with new genetic inventions, which differ markedly from mechanical and chemical inventions that have been the traditional subject matters of agricultural patents. Modern biotechnology inventions, particularly genetic inventions, have become more valuable as an embodiment of information as compared to their physical attributes. The advent of bioinformatics and genetic databases demands a different patent approach, much in tune with the present information age. Although it is not possible to create new IPR every time when a new technology emerges, however, fitting all sorts of inventions in a single set of law is also problematic. The IPR laws are essentially of territorial nature and grant subject matter for national sovereignty. Given the territorial nature of IPR laws, the national/regional laws usually vary and differ in the scope and coverage beyond the basic minimum standard of protectable subjects prescribed by TRIPS especially in the area of patent protection. Even among the USA, European Union, and India, as would be apparent from the above discussions, the scope of patentable subject matter varies from one jurisdiction to another based on the legislative framework, patent practice and interpretation and guidance flowing from the court's precedents.

Agricultural-based countries like India in order to keep pace with the domestic and global requirement of agricultural produce should constantly work on creative developments to meet global challenges and advancements. Given the highly competitive global market with consumer preferences shifting dynamically for better products and advancements to make products better and yet cost effective, it is the IP rights on such advancement that can benefit the creative efforts of persons involves in such research and developmental activities in agricultural sectors. Thus, IPR helps incentivizing research efforts and costs and without IPR rights the future of research and development in agricultural sector may not be viable since anyone who is spending on R&D need to have some form of exclusive rights on the creative/intellectual efforts as otherwise the motivation to create new and advanced products and process will be lost and the society at large will face the consequences of staying with old technology and the entire field of agricultural sector will be affected. The richness of countries having sound agricultural basis/platform like India can be enhanced only through effective utilization of IPR laws and channelizing the intellectual advancements through the strategized geographic domain-based utilities and IPR legal provisions in place.

The IPR therefore facilitates in the following:

(a) to differentiate products and services;
(b) to promote products and services and creates a loyal clientele;
(c) to diversify market strategies to various target groups;
(d) to popularize in foreign countries;
(e) to keep away competitors/copiers.

It is thus important to appreciate and make effective use of IPR in various jurisdictions for facing the global challenges of IP-driven market and dominance, which is expected to the drive the future of all technical field including the agricultural sector as well.

KEYWORDS

- **agriculture**
- **industrial design**
- **IP rights**
- **patent**
- **UPOV**

REFERENCES

1372/DELNP/2006 "*Process for Producing Rose Plant Cell with Modified Color*" dt. 2006/03/13 Suntory Holdings Limited, Japan.

1425/DELNP/2004 Method for modifying plant morphology, biochemistry and physiology comprising expression of plant cytokinin oxidase, dt. 2004/05/26 Inventor & Applicant: Thomas Werner Thomas Schmulling, Germany.

1627/DEL/2009 "Synergistic Insecticidal and larvicidal botanical compositions" dt 2009/08/04 Institute of Pesticide Formulation Technology, India.

1750/MUMNP/2009 dated 18th September, 2009 titled "Process of Producing Tomato Paste" in the name B. V. Nunhems.

2212/MUMNP/2009 "*New Hybrid system for Brassica Napus*" dt 2009/11/27 Applicant: Syngenta Participations Ag.

2407/DELNP/2006 "*Method for Enhancing Stress Tolerance in Plants and Methods Thereof*" dt. 2006/05/01, Applicant: Monsanto Technology LLC, USA.
2668/KOLNP/2008 dated 2nd July, 2008 entitled "*VIP Fragments and Methods of Use*" in the name of Vectus Biosystems Limited.
519/MUMNP/2010 dated 16th March 2010, "*An Article of Manufacture Comprising of a Three–Dimensionally (3D) Cultured Expanded Cell Population of Post Partum Adherent Placental Cells*" Applicant Pluristem LTD, Israel.
5583/DELNP/2010 dt "*Agent for Improving Good Rice Seedling Growth*" applicant: Cosmo Oil Co., Ltd., Japan.
5699/DELNP/2007 "*Transgenic Plants with Enhanced Agronomic Traits*" 2007/07/23; Applicant: Monsanto Technology, LLC, USA.
569 U.S. https://www.supremecourt.gov/opinions/12pdf/12–398_1b7d.pdf. 2013.
9827/DELNP/2007 "*Method of Reducing Mycotoxin Contamination of the Harvest*" dt. 2007/12/18 Applicant: Syngenta Participation Ag, Switzerland.
Application No. 1602/MAS/1998 the patents act, 1970 – Patent Opposition Database.
Application No. 787/MUMNP/2010 Genes for improving salt tolerance and drought tolerance of Plant and the uses thereof dt. 2010/04/19 Applicant: Biotechnology Research Institute, the Chinese Academy of Agricultural Sciences, China.
Application no. 973/MUMNP/2010 Rice non-endosperm tissue expression promoter (OSTSP I) and the use thereof dt. 2010/05/11 Applicant: Rice Research Institute, Anhui Academy Of Agricultural Sciences, 40, Nong Nan Road, Anhui 230031 Hefei, China. Cncnsyngenta Participation Ag schwarzwaldallee 215, Ch-4058 Basel, Switzerland. Chch.
Article 27. 2, 27. 3, TRIPs Agreement.
Article 53 – Exceptions to patentability, The European Patent Convention.
Basmati Case Study. www1.american.edu/TED/basmati.htm.
Basmati Patent US5663484, RiceTec.in.
Biological diversity act–2002–National Biodiversity Authority, www.nbaindia.org/.
Darjeeling Tea – A Geographical Indication, Tea Board of India. www.teaboard.gov.in/pdf/policy/geographical_indication_for_darjeeling_tea.doc.
Decision: supremecourtofindia.nic.in/outtoday/patent.pdf Supreme court of India Order: Civil Appeal Nos. 2706–2716 OF 2013(Arising Out of SLP(C) Nos. 20539–20549 of 2009).
Decision of the enlarged board of appeal 25th March 2015 G2/13.
Dimminaco, A. G., (2001). Versus controller of patents and designs & others in the high court at Calcutta judgment On: 15. 01.
EP 0 728 048 B1 Waste treatment proprietor: Geohess (UK) Limited Stowe, Buckingham.
EP 1 069819: Method for selective increase of the anti-carcinogenic glucosinolates in Brassica napus: proprietor: Plant Bioscience Limited, Norwich.
EP 1 211 926: Method for Breeding Tomatoes having reduces water content and product of the method : Proprietor :State of Israel Ministry of Agriculture, 26.11.2003.
EP 1656150 "Use of bacilli bacteria for the prevention of vertebral compression syndrome in salmonids" dt 23. 11. Institut Francais De Recherche Pour L'exploitation De La Mer (Ifremer) 92138 Issy-les-Moulineaux Cedex (FR) & Institut National De La Recherche Agronomique 75341 Paris Cédéx 07 (FR), 2006.

European Union (EU) Directive 98/44/EC of the European Parliament and of the Council of 6 July 1998 on the legal protection of biotechnological inventions.
Feldman, C., et al., (2007). Lessons from the commercialization of the Cohen-*Bo*yer patents: The Stanford University Licensing Program, Handbook of Best Practices |1797, Chapter 17, 22.
GI – Intellectual Property India. www.ipindia.nic.in/gi.htm.
http://ipindiaservices.gov.in/GirPublic/ViewApplicationDetails.
http://www.ipindia.nic.in/history-of-indian-patent-system.htm).
http://www.ipindia.nic.in/writereaddata/Portal/News/159_1_115-public-notice-02july2014.pd. 17 a. Decision of the enlarged board of appeal 25th March 2015 G2/12.
http://www.plantauthority.gov.in/.
https://en.wikipedia.org/wiki/TRIPS_Agreement.
https://supreme.justia.com/cases/federal/us/447/303/case.html Diamond v. Chakrabarty, 447 U. S. 303 (1980).
https://www.epo.org/law-practice/legal-texts/html/guidelines/e/f_iv_4_16.htm.
Indian Patent Act, (1970), Chapter II: Inventions Not Patentable, pp. 8.
In PASS – Indian patent advanced search system ipindiaservices.gov.in/publicsearch/.
Introduction to the TRIPS Agreement – World Trade Organization. https://www.wto.org/english/tratop_e/trips_e/ta_docs_e/modules1_e.pdf.
IPAB order dt. 08/07/2013: OA/2/2012/PT/DEL and M. P. NO. 35 & 36/2013 IN OA/2/2012/PT/DEL.
List of Geographical Indications in India – Wikipedia. https://en.wikipedia.org/wiki/List_of_Geographical_Indications_in_India.
MPEP Patent Laws [PDF] – United States Patent and Trademark Office; www.uspto.gov/patent/laws-and-regulations/examination-policy/2014-interim-guidance-subject-matter- ligibility-0.
No-patents-on-seeds.org/en/./european-patent-office-revokes-monsanto-patent-melon.
Ref: Guidelines for Examination of Biotechnology Applications for Patents.
Seed Act, (1966) seednet.gov.in/PDFFILES/Seed_Act_1966.pdf.
T 0416/01 (Method for controlling fungi on plants/thermo trilogy corporation) of 8.3.2005 https://www.epo.org/.
T 1054/02 of 6.5. Termination of the opposition proceedings with revocation of patent No. 0 728 048, 2004.
The Patents Act, (1970), as amended by Act No. 15 of 2005 – WIPO.
United States Patent and Trademark Office, http://www.uspto.gov.
US 4237224 (1980) "Process for producing biologically functional molecular chimeras."
US4259444, Microorganisms having multiple compatible degradative energy-generating plasmids and preparation thereof 31. 03. 1981; Original Assignee General Electric Company.
WIPO website, summary of the Budapest treaty on the international recognition of the deposit of microorganisms for the purposes of patent procedure, (1977), www.wipo.int/export/sites/www/treaties/en/registration/budapest/./introduction.pdf].
www.ipindia.nic.in/acts-rules-tm.htm.
www.ipindia.nic.in/designs.htm.
www.plantauthority.gov.in_List_of_Cerificates.pdf.

CHAPTER 2

BIODIVERSITY CONSERVATION AND AGRICULTURAL INTENSIFICATION IN INDIA THROUGH INTEGRATION OF IPR

SUJIT KUMAR[1] and SEWETA SRIVASTAVA[2]

[1] *U.P. Council of Agricultural Research, Lucknow–226 010 (U.P.), India*

[2] *School of Agriculture, Lovely Professional University, Phagwara–144 411, Punjab, India, E-mail: seweta.21896@lpu.co.in*

CONTENTS

ABSTRACT

Integrating the conservation of biodiversity with agricultural intensification is increasingly through intellectual property rights (IPRs) recognized as a leading priority of sustainability and food security amid global environmental and socioeconomic change. At the genetic, species, and farming systems levels, biodiversity provides valuable ecosystems services and functions for agricultural production. The successful protection and deployment of biodiversity hinges on a favorable policy environment and on agricultural research and extension activities that stress farmer participation and greater sensitivity to the off-site impacts of agriculture. However, in spite of the laws, we are rapidly losing our diversity that poses threat to our ecological balance and also contributes to climatic changes. We need to know where the lacuna is. This article highlights key principles, policies, and practices for the sustainable use, conservation, and enhancement of agro-biodiversity for sustaining agriculture.

2.1 INTRODUCTION

Biodiversity is at the heart of the intricate web of life on the Earth and the processes essential to survival. Our planet's biological resources are not only shaped by natural evolutionary processes but also increasingly transformed by anthropogenic activity, population pressures, and globalizing tendencies. When human activity threatens these resources, or the complex ecosystems of which they are a part, it poses potential risks to millions of people whose livelihoods, health, and well-being are sustained by them. The increasingly complex global health challenges that we face, including poverty, malnutrition, infectious diseases, and the growing burden of non-communicable diseases (NCDs), are more intimately tied than ever to the complex interactions between ecosystems, people, and socioeconomic processes. India is a country of rich biological diversity, having more than 91,200 species of animals and 45,500 species of plants in its 10 bio-geographic regions. Besides, it is recognized as one of the eight Vavilovian centers of origin and diversity of crop plants, with more than 300 wild ancestors and close relatives of cultivated plants, which are still evolving under natural conditions. India is also a vast repository of traditional knowledge associated with biological resources. India ranks among the top 10 species-rich nations and shows high endemism. It has four global biodiversity hot spots. The varied edaphic, climatic, and topographic conditions and years of geological stability have resulted in a wide range of ecosystems and habitats in India. Unfortunately, as elsewhere on the Earth, Indian biodiversity is also threatened with destruction due to population pressures and ill-conceived developmental activities. The Government of India became aware of the situation and created wildlife sanctuaries, national parks, and biosphere reserves for in situ conservation of biodiversity and various scientific organizations (gene banks) for ex situ conservation (Chauhan, 2014).

The dual challenges of biodiversity loss and rising global health burdens are not only multifaceted and complex but also transcend sectoral, disciplinary, and cultural boundaries, and demand far-reaching, coherent and collaborative solutions. One of the widely acknowledged shortcomings of the millennium development goals (MDGs) and targets (the precursors of the sustainable development goals (SDGs)) was the lack of cross-sectoral integration among social, economic, and environmental goals, targets, and priorities (Haines et al., 2012).

The World Health Organization (WHO) and the Secretariat of the Convention on Biological Diversity (CBD) are working together to address these challenges. This state of knowledge review assembles expertise and insights from numerous researchers, practitioners, policymakers, and experts from the fields of biodiversity conservation, public health, agriculture, nutrition, epidemiology, immunology, and others to do the following:

- Provide an overview of the scientific evidence for linkages between biodiversity and human health in a number of key thematic areas;

- Contribute to a broader understanding of the importance of biodiversity to human health in the evolving context of the SDGs and post-2015 Development Agenda, as well as the Strategic Plan for Biodiversity 2011–2020 (see https://www.cbd.int/sp/);
- Facilitate cross-sectoral, interdisciplinary, and trans-disciplinary approaches to health and biodiversity conservation, and promote co-operation between different sectors and actors in an effort to mainstream biodiversity in national health strategies and mainstream health in biodiversity strategies;
- Provide some of the basic tools necessary to investigate how biodiversity may influence health status or health outcomes, for given projects, policies, or plans at varying levels (i.e., from community to the national, regional, and global levels).

This chapter is aimed primarily at policymakers, practitioners, and researchers working in the fields of biodiversity conservation, public health, academicians, agricultural development, and other relevant sectors. Its findings suggested that greater interdisciplinary and cross-sectoral collaboration is essential for the development of more coordinated and coherent policies aimed at addressing the tripartite challenge of biodiversity loss and the global burden of ill-health and development. Interdisciplinary scientific investigation and approaches are critical to meet these challenges. The full involvement of all segments of society, including local communities, will also be needed for transition toward a new era of sustainable development.

2.2 BIODIVERSITY

Biological diversity, most commonly used in its contracted form, biodiversity, is the term that describes the variety of life on the Earth, including animals, plants, and microbial species. It has been estimated that there are some 8.7 million eukaryotic species on the Earth, of which some of the 25% (2.2. million) are marine, and most of them are yet to be discovered (Mora et al., 2011). Biodiversity not only refers to the multitude of species on the Earth but also consists of the specific genetic variations and traits within species (such as different crop varieties), and the assemblage of these species within ecosystems that characterize agricultural and other landscapes such as forests, wetlands, grasslands, deserts, lakes, oceans, and rivers. Each ecosystem comprises living beings that interact with one another and with the air, water, environment, and soil around them. These multiple interconnections within and between the ecosystems forms the web of life, of which human being are an integral part and upon which they depend for their very survival. It is the combination of these life forms and their interactions with one another, and with the surrounding environment, that makes human life on the Earth possible (CBD, 2006).

2.3 THE BIOLOGICAL DIVERSITY ACT 2002

It is an Act of the Government of India for the preservation of biological diversity in India. The Act was enacted to meet the obligations under CBD, to which India is a member. Biodiversity has been defined under Section 2(b) of the Act as the variability among living organisms from all sources and the ecological complexes of which they are part, and includes diversity within species or between species and of eco-systems. The Act also defines biological resources as plants, animals, and micro-organisms or parts thereof, their genetic material and by-products (excluding value-added products) with actual or potential use or value but does not include human genetic material.

The Act covers conservation, use of biological resources, and associated knowledge occurring in India for commercial or research purposes or for the purposes of bio-survey and bio-utilization. It provides a framework for access to biological resources and sharing the benefits arising out of such access and use. The Act also includes in its ambit the transfer of research results and application for IPRs relating to Indian biological resources. The major components of the act are described in Figure 2.1.

The main objectives of the Biological Diversity Act 2002 are:

1. The conservation of biodiversity;
2. The sustainable use of biological resources;
3. Equity in sharing benefits from such use of resources.

The Act covers foreigners, non-resident Indians, corporate bodies, association, or organization that is either not incorporated in India or incorporated in India with non-Indian participation in its share capital or man-

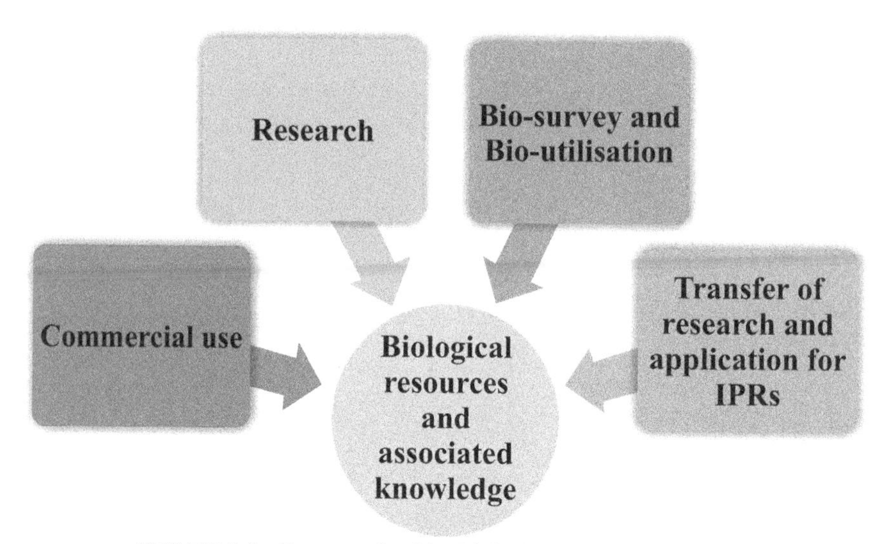

FIGURE 2.1 Framework of the Biological Diversity Act 2002.

agement. These individuals or entities require the approval of the National Biodiversity Authority when they use biological resources and associated knowledge occurring in India for commercial or research purposes or for the purposes of bio-survey or bio-utilization (Figure 2.2).

Indians and Indian institutions do not require the approval of the National Biodiversity Authority when they engage in the abovementioned activities. However, they would need to inform the State Biodiversity Boards prior to undertaking such activities. Any commercial application related to use of biological resources should, however, be approved by the Authority.

The Act excludes Indian biological resources that are normally traded as commodities and for no other purpose. The Act also excludes traditional uses of Indian biological resources and associated knowledge and when they are used in collaborative research projects between Indian and foreign institutions with the approval of the central government (Figure 2.3).

2.4 KEY PROVISIONS OF THE BIOLOGICAL DIVERSITY ACT 2002

1. Prohibition on transfer of Indian genetic material outside the country, without specific approval of the Indian Government;

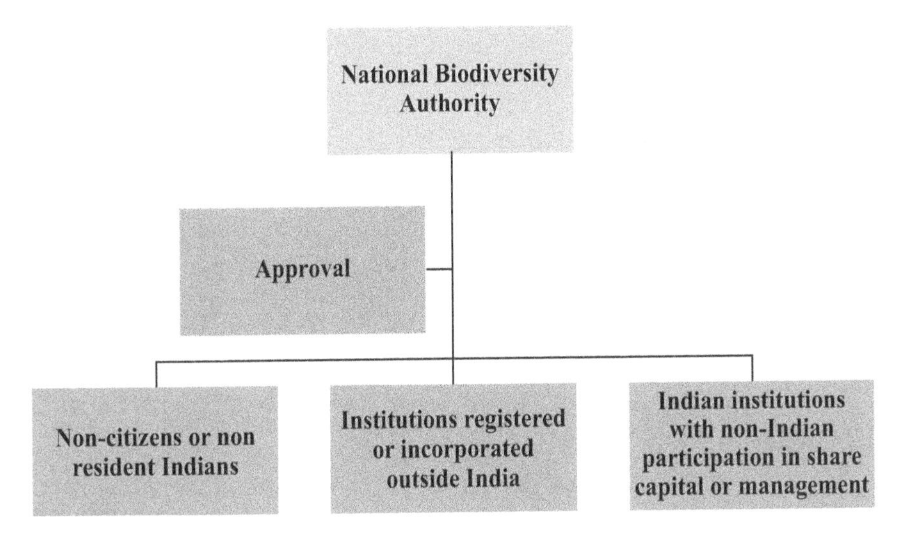

FIGURE 2.2 Inclusion of the Biological Diversity Act 2002.

2. Prohibition on anyone claiming an intellectual property right (IPR), such as a patent, over biodiversity or related knowledge, without permission of the Indian Government;

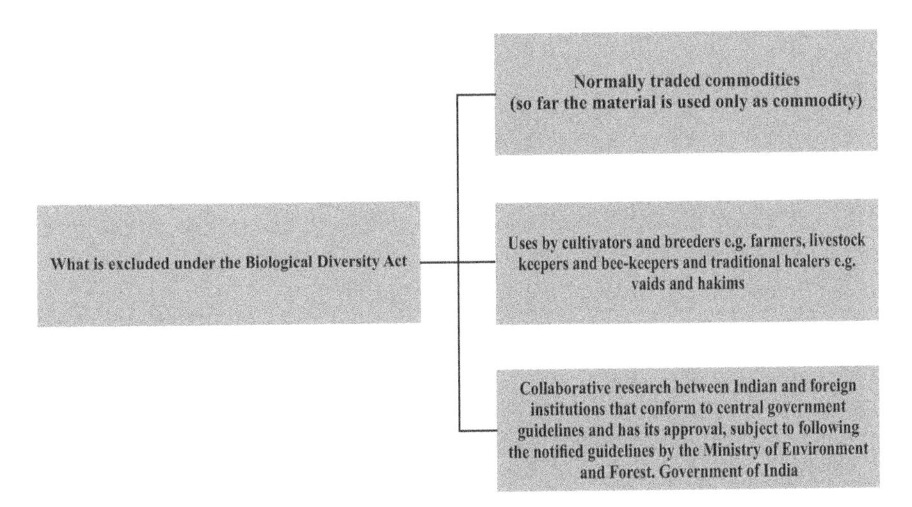

FIGURE 2.3 Exclusion of the Biological Diversity Act 2002.

3. Regulation of collection and use of biodiversity by Indian nationals, while exempting local communities from such restrictions;
4. Measures for sharing of benefits from the use of biodiversity, including transfer of technology, monetary returns, joint Research & Development, joint IPR ownership, etc.;
5. Measures to conserve and sustainably use biological resources, including habitat and species protection, environmental impact assessments (EIAS) of projects, integration of biodiversity into the plans, program, and policies of various departments/sectors;
6. Provisions for local communities to have a say in the use of their resources and knowledge, and to charge fees for this;
7. Protection of indigenous or traditional knowledge, through appropriate laws or other measures such as registration of such knowledge;
8. Regulation of the use of genetically modified organisms;
9. Setting up of national, state, and local biodiversity funds to be used to support conservation and benefit-sharing;
10. Setting up of Biodiversity Management Committees (BMC) at the local village level, State Biodiversity Boards (SBB) at the state level, and a National Biodiversity Authority of India (NBA).

2.5 WEAKNESS OF SOME OF THE PROVISIONS

1. It exempts those plants that are registered under the Protection of Plant Varieties and Farmers Rights (PVP & FR) Act 2001. This Act provides corporations and scientists who are breeding new varieties of crops to gain intellectual property rights. Such an exemption means that the progressive provisions listed above, many of which are absent from the PVP & FR Act would not apply to plant varieties registered under PVP & FR Act.
2. It does not provide citizens the power to directly approach the courts; such power is restricted to an appeal in the High Court against any order by the NBA or the SBB.
3. It is unnecessarily soft on Indian corporate and other entities, requiring only "prior intimation" to a SBB for the commercial use

of bioresources, rather than permission from the NBA as in the case of foreigners. This is unjustified, given that Indians (especially industrial corporations) are not necessarily any more responsible toward the environment or toward local communities; also, some Indian companies could just be local fronts for foreign enterprises.

4. It does not fully empower local communities to protect their resources and knowledge from being misused or to generate benefits (except charging collection fees). It has very weak or no representation of local community members on the SBB or NBA.

2.6 THE ROLE OF STATE GOVERNMENTS

The power of declaring a biodiversity heritage sites lies with the state government (Article 37 of the Act). It is important that the heritage sites should be designated only after consultation and moreover consent of the affected communities. Further, these should be in the control/management of local communities, and the provision for compensation made in the state biodiversity fund (see Section 32) be applied only where there is a mutually agreement to dislocation/curbing of rights. Several organizations and people feel that the basic framework of the Act is problematic, as it accepts intellectual property rights on biodiversity, could be used to further commercialize biodiversity, and does not truly empower communities. Others feel that the Act provides some potential for checking biopiracy, achieving conservation, and facilitating community action. They stress that a combination of strong rules, and amendments related to the above points, would help to strengthen the potential of Biodiversity Act.

2.7 BIODIVERSITY RULES, 2004

There was hope that Rules under the Act would strengthen the provisions on conservation, sustainable use, and equity. Unfortunately, that hope was shattered when the government notified the Biological Diversity Rules 2004 on April 15. The Biodiversity Rules are the executive orders made by the government in order to carry out the purposes of the Act (Section

62). Every rule made under this Act was placed in the Parliament for a period of 30 days, and the houses approved the rules with amendments (sec 62(3)). The rules among other things outline the procedures to be followed for access to biological resources (wild plants and animals, crops, medicinal plants, livestock, etc.), their commercial utilization, transfer of rights of research, and IPRs related to biodiversity. From the point of view of local communities, it is important to understand the process of allowing access/utilization of bio-resources and also the role of communities. Some provisions are directly relevant to local communities, the most critical of them being the Biodiversity Management Committee (BMC).

Section 41 of the Act states: Every local body shall constitute a Biodiversity Management Committee within its area for the purpose of promoting conservation, sustainable use, and documentation of biological diversity including preservation of habitats, conservation of land races, folk varieties and cultivars, domesticated stocks and breeds of animals and microorganisms and chronicling of knowledge related to biological diversity. Under the Biodiversity Rule, Sec 22 expands on constitution and role of Biodiversity Management Committees, and states: issues in the Parliament 2(h) local bodies means Panchayats and Municipalities, by whatever name called, within the meaning of clause (1) of article 243B and clause (1) of article 243Q of the Constitution and in the absence of any Panchayats or Municipalities, institutions of self-government constituted under any other provision of the Constitution or any Central Act or State Act.

2.7.1 EVERY LOCAL BODY SHALL CONSTITUTE A BIODIVERSITY MANAGEMENT COMMITTEE (BMCS) WITHIN ITS AREA OF JURISDICTION

The main function of the BMC is to prepare People's Biodiversity Register (PBR) in consultation with local people (this is a comedown from the broader role envisaged in Sec 41) of the Act. The Register is supposed to contain comprehensive information on availability and knowledge of local biological resources, their medicinal or any other use, or any other traditional knowledge associated with them. The other functions of the BMC are to advice on any matter referred to it by the SBB or Authority for granting approval and to maintain data about the local vaids and practitioners using the biological resources. Therefore, the role for BMCs defined in the Biodiversity Rules are a complete comedown from what was envisaged in the Biodiversity Act, which itself had its own set of problems.

2.7.2 CONSTITUTION OF THE BIODIVERSITY MANAGEMENT COMMITTEES (BMC)

a. The definition of local body is problematic, as it leaves out gram sabha or other village assemblies. As the local body has to appoint/select the BMC, the political affiliation and relationship between a village and the panchayat body will play an important role in the constitution and functioning of the BMC.
b. The process of local body constituting BMC is by nomination. Rules 22(2) & (3) expressly mention that the members will be NOMINATED by the local body and the Chairperson will be ELECTED by the committee. Then, the BMC could become another power center and might not actually function to conserve biodiversity or protect community rights.

2.7.3 FOCUS OF WORK AND FUNCTIONING

a. The Act clearly spells out a list of functions for the BMC, among which are promoting conservation and maintaining PBR. The Rule dilutes this and states that the main role is to merely maintain PBR.
b. Peoples Biodiversity Register (PBR): The PBR is a document that records the diversity of species of flora, fauna, crops, livestock etc. As on date, there is no legal protection available for the knowledge recorded in the PBR. This is problematic when it comes to the question of access to this document. Even though communities create and maintain a database of their resources of knowledge, there is no requirement that their consent would sought when it comes to accessing the information in the PBRs. Although Rule 17 says local bodies will be consulted before

approval for access to bioresources is given, the definition of "consult" is not clear, and in many cases, it might remain a mere formality.

c. Though the Act clearly has spelt out criteria for rejecting applications, it has not listed community consent as one of them. Rule 7 is clearly biased, as it gives BMC only an advisory role in the grant of approvals.

2.8 BIODIVERSITY ACT/RULES AND INTELLECTUAL PROPERTY RIGHTS

The biodiversity legislation provides for a regulatory system by which access to knowledge relating to biodiversity can be granted. Providing for an approval procedure for a patent or any other IPR based on any Indian biological material and knowledge is seen by several groups campaigning against patents on life as a significant departure from the earlier stance of the Government of India. The Act does not prohibit IPRs, and therefore, the criticism is that it facilitates the privatization of India's traditional knowledge. The Act only forbids an application for any IPR in or outside India without prior approval of the NBA (Section 6). The NBA may either allow or disallow an application for a patent or any other IPRs. Neither the procedure in the relevant Rule 18 nor the Form III for seeking such approval factors in consultation of communities. On occurrence of an instance of biopiracy, the NBA is empowered by the Act, to take any necessary action to oppose the grant of IPR in any country outside India on behalf of the Government of India [Section 18(4)]. In the absence of a globally agreed single forum wherein such cases can be challenged, the NBA may have to only engage in fire-fighting at different patent and or trademark offices overseas. Indian trade negotiators have at international forum agreed that patents will be allowed on such resources or knowledge only if there is:

a. Disclosure of source and country of origin of the biological resource and of the traditional knowledge used in the invention;
b. Disclosure of evidence of prior informed consent;
c. Disclosure of evidence of benefit sharing IPRs.

2.9 IN THE CONTEXT OF PROTECTION OF PLANT VARIETIES AND FARMERS RIGHT ACT

The IPR provisions in the Biodiversity Act must also be seen in the light of the growing pro-IPR trend of the Government of India, more visible in other IPR-related laws and policies such as: the Protection of Plant Varieties and Farmer's Rights Act 2001, which introduces plant breeders' rights Amendments in the Patent Act 1970 toward compliancy of WTO TRIPS' standards. An IPR sought under the PVP law does not come under the purview of the Biodiversity Act; in other words, a person seeking a plant breeder right does not require approval of the NBA. The PVP Authority is only to keep the NBA informed of such grant of rights. Therefore, the entire three legislations (Biodiversity Act, Protection of Plant Varieties and Farmer's Rights Act, and Patent Act) move in tandem toward a pro-IPR regime, and in that sense are not "incompatible." So, even though an international convention like the CBD (Article16.5) states that IPRs must not conflict with the conservation and sustainable use of biodiversity, the biodiversity law is apparently based on the premise that IPRs and biodiversity conservation are not antithetical. It is important to note that is only after India became a signatory to this convention, the process of drafting the legislation begun in India. This is how the IPR philosophy or rather politics (contained in the Patents & PVP Legislation) has even corrupted a supposedly conservation-oriented legislation.

2.10 CONVENTION ON BIOLOGICAL DIVERSITY (CBD)

The Convention on Biological Diversity (CBD) entered into force on 29 December 1993. It has three main objectives:

1. The conservation of biological diversity.
2. The sustainable use of the components of biological diversity.
3. The fair and equitable sharing of the benefits arising out of the utilization of genetic resources.

The CBD defines biodiversity as: the variability among living organisms from all sources including, inter alia, terrestrial, marine and other aquatic ecosystems and the ecological complexes of which they are part; this includes diversity within species, between species, and of ecosystems. Biodiversity encompasses much more than the variety of life on the Earth; it also includes biotic community structure, the habitats in which communities live, and the variability within and among them. Thus, biodiversity extends beyond the simple measurement of species numbers to include the complex network of interactions and biological structures that sustain ecosystems (McCann, 2007; Maclaurin and Sterelny, 2008). Although "species richness" is one of biodiversity's key components, the two terms are not synonymous.

The widely accepted definition of biodiversity adopted by the CBD is flexible, inclusive, and reflective of the levels and complexities of biotic and abiotic interactions. It recognizes levels of variability within species, between species, and within and between ecosystems as integral to the ecological processes of which they are a part (Mace et al., 2012). It is also understood that variability manifests itself differently at various temporal and spatial scales (Nelson et al., 2009; Thompson et al., 2009).

The scope of the Convention is broader still; its objectives—the conservation of biological diversity, the sustainable use of its components and the fair and equitable sharing of the benefits arising out of the utilization of genetic resources—indicate an interest in the components of biodiversity (including individual species) and genetic resources.

2.11 BIODIVERSITY, ECOSYSTEM FUNCTIONS, AND SERVICES

Scientific knowledge of the impacts of biodiversity loss on ecosystem functioning has increased considerably in the past two decades (Tilman et al., 1997; Loreau et al., 2001; Naeem and Wright, 2003; Cardinale et al., 2012; Balvanera et al., 2014) as well as corresponding knowledge of its implications for public health (Myers et al., 2013). In this section, we summarize key elements of the relationship among biodiversity, ecosystems, and ecosystem functioning; its connection to ecosystem services; and the components that influence the quantity, quality, and reliability of ecosystem services, and that contribute to ecosystem resilience.

There is strong evidence of the relationship between biodiversity and ecosystem functioning and, in some cases, we can directly link this to the ecosystem services necessary to sustain human health (Loreau et al., 2001;

Balvanera et al., 2006; Cardinale et al., 2012; Balvanera et al., 2014). In other cases, we do not have complete evidence of this relationship (Schwartz et al., 2000; Cardinale et al., 2012). While there is broad consensus within the scientific community on several aspects of the relationship among biodiversity, ecosystem functioning, and the consequences of its loss on the ability of ecosystems to provide services, the full range of impacts of biodiversity loss on ecosystem functioning is not fully understood (Hooper et al., 2005; Reiss et al., 2009).

Twenty years of work on the relationship between biodiversity and ecosystem functioning have generated a number of controversies and spurred efforts to develop scientific consensus. Cardinale et al. (2012) conclude that diverse communities tend to be more productive both because they contain key species that have a large influence on productivity, and because differences in functional traits among organisms increase the total capture of resources (light, water). Thus, biodiversity loss reduces the efficiency by which ecological communities capture biologically essential resources, produce biomass, and decompose and recycle biologically essential nutrients. They report that the impact of biodiversity on any single ecosystem process is nonlinear and saturating, such that change accelerates as biodiversity loss increases. They also point to mounting evidence that biodiversity increases the stability of ecosystem functions through time.

2.12 BIO-GEOGRAPHICAL ZONES OF INDIA

India represents (i) Two Realms—the Himalayan region represented by Palaearctic Realm and the rest of the sub-continent represented by Malayan Realm; (ii) Five Biomes, for example, Tropical Humid Forests; Tropical Dry Deciduous Forests (including Monsoon Forests); Warm Deserts and Semi-deserts; Coniferous Forests; Alpine Meadows; and (iii) 10 bio-geographic zones (Chauhan, 2008) and 27 bio-geographic provinces (Table 2.1).

2.12.1 ZONE: TRANS-HIMALAYA

The Trans-Himalaya Zone covers an area of 186,200 square km in the cold and arid regions with sparse alpine steppe vegetation and several endemic

TABLE 2.1 Bio-Geographical Zones of India

Sl. No.	Bio-geographical Zones	Bio-geographical Provinces	Percent of Geographical Area of India
1.	Trans Himalaya	1A: Himalaya-Ladakh Mountains	3.3
		1B: Himalaya -Tibetan Plateau	2.2
		1C: Trans-Himalaya Sikkim	<0.1
2.	The Himalaya	2A: Himalaya-North West Himalaya	2.1
		2B: Himalaya-West Himalaya	1.6
		2C: Himalaya-Central Himalaya	0.2
		2D: Himalaya-East Himalaya	2.5
3.	The Indian Desert	3A: Desert-Thar	5.4
		3B: Desert-Kutch	1.1
4.	The Semi Arid	4A: Semi-Arid-Punjab Plains	3.7
		4B: Semi-Arid-Gujarat Rajputana	12.9
5.	The Western Ghats	5A: Western Ghats-Malabar Plains	2.0
		5B: Western Ghats -Western Ghats Mountains	2.0
6.	The Deccan Peninsula	6A: Deccan Peninsular-Central Highlands	7.3
		6B: Deccan Peninsular-Chotta Nagpur	5.4
		6C: Deccan Peninsular-Eastern Highlands	6.3
		6D: Deccan Peninsular-Central Plateau	12.5
		6E: Deccan Peninsular-Deccan South	10.4
7.	The Gangetic Plains	7A: Gangetic Plain-Upper Gangetic Plains	6.3
		7B: Gangetic Plain-Lower Gangetic Plains	4.5
8.	The Coasts	8A: Coasts-West Coast	0.6
		8B: Coasts-East Coast	1.9
		8C: Coasts-Lakshadweep	<0.1
9.	Northeast India	9A: North-East-Brahamputra Valley 9B: North-East-North East Hills	2.0 3.2
10.	Islands	10A: Islands-Andamans	0.2
		10B: Islands-Nicobars	0.1

species. It is home to communities of wild sheep and goats, the urial, ibex, wild yak and Tibetan ass, gazelle, and antelope. Among the carnivores are the snow leopard, Tibetan wolf, and the endemic pallas cat and smaller animals such as the marbled cat, pika, and marmot. The brackish water lakes and marshes have a good variety of avifauna, the most spectacular of which is the black-necked crane. The zone has two protected areas. This zone is virtually treeless, except for cultivated varieties of *Populus* and *Salix* along the major water courses (Kala and Mathur, 2002).

2.12.2 ZONE: THE HIMALAYA

This zone covers an area of about 236,000 square km in the Himalayas. The Himalayas have attained a unique personality owing to their high altitude, steep gradient, and rich temperate flora. The forests are very dense with extensive growth of grass and evergreen tall trees. Oak, chestnut, conifer, ash, pine, and deodar are abundant in Himalayas. There is no vegetation above the snowline. It displays a wide altitudinal range and is among the richest zones in terms of species and habitat diversity—the sambar, muntjac, wild boar in the subtropical foothills; the musk deer, serow, goral, tahr, kokla, and pheasant in the temperate, sub-alpine regions; and the bharal, snow leopard, brown bear, and snowcock in the alpine region. The zone has 56 protected areas. There are more endangered species in the Himalaya than anywhere else in India. The Sikkim stag may already have become extinct. The tahr, markhor, and western tragopan may be facing extinction.

Ceylon ironwood (*Mesua ferrea*) is found on porous soils at elevations between 600 and 2,400 feet (180 and 720 meters); bamboos grow on steep slopes; oaks (genus *Quercus*) and Indian horse chestnuts (*Aesculus indica*) grow on the lithosol (shallow soils consisting of imperfectly weathered rock fragments), covering sandstones from Arunachal Pradesh westward to central Nepal at elevations from 3,600 to 5,700 feet (1,100 to 1,700 meters). Alder trees (genus *Alnus*) are found along the watercourses on the steeper slopes. At higher elevations, those species give way to mountain forests in which the typical evergreen is the Himalayan screw pine (*Pandanus furcatus*). Besides those trees, some 4,000 species of flowering plants, of which 20 are palms, are estimated to occur in the eastern Himalayas. Juniper (genus *Juniperus*) is widespread, especially on sunny sites, steep, and rocky slopes, and drier areas. Rhododendron occurs everywhere but is more abundant in the wetter parts of the eastern Himalayas, where it grows in all sizes from trees to low shrubs. Mosses and lichens grow in shaded areas at lower levels in the alpine zone where the humidity is high; flowering plants are found at high elevations.

2.12.3 ZONE: THE INDIAN DESERT

This third zone is a highly fragile ecosystem and its biological richness may be lost very rapidly. The zone covers an area of 225,000 square km, of which about 89 square km is protected. The wild ass, a distinct subspecies, is restricted to the Rann of Kutch. This is also the only breeding site for flamingoes on the Indian subcontinent. It is home to the desert fox, desert cat, houbara, bustard, and to some sandgrouse species. Other species are the chinkara, blackbuck, wolf, caracal, and great Indian bustard.

The natural vegetation of this dry area is classed as North-Western thorn scrub forest occurring in small clumps scattered more or less openly (Champion and Seth, 1968; Negi, 1996). Density and size of patches increase from west to east following the increase in rainfall. The natural vegetation of the Thar Desert is composed of the following tree, shrub, and herb species (Kaul, 1970):

- **Trees and Shrubs:** *Acacia jacquemontii, Balanites roxburghii, Ziziphus zizyphus, Ziziphus nummularia, Calotropis procera, Suaeda fruticosa, Crotalaria burhia, Aerva javanica, Clerodendrum multiflorum, Leptadenia pyrotechnica, Lycium barbarum, Grewia tenax, Commiphora mukul, Euphorbia neriifolia, Cordia sinensis, Maytenus emarginata, Capparis decidua, Mimosa hamata.*
- **Herbs and Grasses:** *Ochthochloa compressa, Dactyloctenium scindicum, Cenchrus biflorus, Cenchrus setigerus, Lasiurus scindicus, Cynodon dactylon, Panicum turgidum, Panicum antidotale, Dichanthium annulatum, Sporobolus marginatus, Saccharum spontaneum, Cenchrus ciliaris, Desmostachya bipinnata, Eragrostis* species, *Ergamopagan* species, *Phragmites* species, *Tribulus terrestris, Typha* species, *Sorghum halepense, Citrullus colocynthis.*

The endemic floral species include *Calligonum polygonoides, Prosopis cineraria, Acacia nilotica, Tamarix aphylla,* and *Cenchrus biflorus* (Khan and Frost, 2001).

2.12.4 ZONE: THE SEMIARID

Adjoining the desert are the semi-arid areas, a transitional zone between the desert and the denser forests of the Western Ghats. Spread over 508,000 square km in the semi-arid regions, zone has two major tiger reserves. There are 52 protected areas covering 11,675 square km. The Gir lion, one of the very few endemic species in this zone, now needs a second home. Birds, jackals, leopards, eagles, snakes, fox, and buffaloes are found in this region. The natural vegetation is thorn forest. This region is characterized by discontinuous vegetation cover with open areas of bare soil and soil-water deficit throughout the year. Thorny shrubs, grasses, and some bamboos are present in some regions. A few species of xerophytic herbs and some ephemeral herbs are found in this semi-arid tract.

2.12.5 ZONE: THE WESTERN GHATS

A 1,500-km-long mountain range with a wet western face and a dry eastern slope, the Western Ghats cover an area of about 159,000 square km. The Western Ghats are among the 25 biodiversity hot spots recognized globally. This zone consists of a diversity of forests, from evergreen to dry deciduous. The richest of India's evergreen forests are located here. It is a continually expanding "genetic storehouse" of India. The Western Ghats cover only 5% of India's land surface but contain more than a quarter (about 4,000) of the country's plant species. About 1,800 of these species are endemic, many highly localized and extremely vulnerable due to increasing habitat destruction. Rice cultivation in the fertile valley proceeded gardens of early commercial crops like areca nut and pepper. Expansion of traditional agriculture and the spread of particularly rubber, tea, coffee, and forest tree plantations would have wiped out large pockets of primary forests in valleys. This zone also has viable populations of most of the vertebrate species found in peninsular India, in addition to endemic species like the Nilgiri langur, the lion-tailed macaque, the Nilgiri tahr, and the Malabar grey hornbill. Most of the amphibian species here too are endemic. The Travancore tortoise and cane turtle are restricted to small areas of the central Western Ghats. There are 44 protected areas covering 15,935 square km. The Western Ghats are well known for harboring 14 endemic species of caecilians (i.e., legless amphibians) out of 15 recorded from the region so far.

2.12.6 ZONE: THE DECCAN PENINSULA

This zone covers 1,421,000 square km—about 43% of India's total land surface. Most of India's protected areas are in this zone. The highlands of the plateau are covered with different types of forests, which provide a large variety of forest products. Most wildlife species—the tiger, leopard, sloth bear, gaur, sambar, chital, chowsingha, wild boar, etc.—are widespread throughout the whole zone. There are small relict populations of elephant, wild buffalo, and barasingha. There are about 115 protected areas covering 4,610 square km.

2.12.7 ZONE: THE GANGETIC PLAIN

Centuries ago, this zone had rich vegetal cover and diverse wildlife, but both are now depleted with the extension of agriculture. The plain supports some of the highest population densities depending upon purely agro-based economy in some of these areas. The trees belonging to these forests are teak, sal, shisham, mahua, khair, etc. The elephant, barasingha, blackbuck, gazelle, rhino, and Bengal florican, which used to be numerous, have only relict populations surviving. In the many wetlands, lakes and swamps, the waterfowl community (partly migratory) is exceptionally dense. Crocodile and freshwater turtle populations are also quite good. Spread over 359,000 square km, the zone has 25 protected areas.

2.12.8 ZONE: THE COASTS

This zone comprises mangrove vegetation and is biologically rich. Extensive deltas of the Godavari, Krishna, and Kaveri are the characteristic features of this coast. Mangrove vegetation is characteristic of estuarine tracts along the coast, for instance, at Ratnagiri in Maharashtra. Larger parts of the coastal plains are covered by fertile soils on which different crops are grown. Rice is the main crop of these areas. Coconut trees grow all along the coast. Animal species include dugong and humpback whale, inshore dolphin, marine and estuarine turtles, and estuarine and saltwater crocodile. Avifauna includes oceanic visitors. The Sunderbans Sanctuary on the east coast is a tiger reserve with Indian's highest population of tigers. Together with Bangladesh, this is one of the world's largest protected mangrove ecosystems.

2.12.9 ZONE: NORTHEAST INDIA

This ninth zone is the biogeographical gateway for much of India's flora and fauna. This zone is one of the richest in biological resources, both endemic and others. The Brahmaputra valley contains extensive areas of natural vegetation—swamps, grasslands, and fringe forests. Northeast India is one of the richest flora regions in the country. It has several species of orchids, bamboos, ferns, and other plants. Here, the wild relatives of cultivated plants such as banana, mango, citrus, and pepper can be grown. The elephant, rhinoceros, buffalo, swamp deer, hog deer, pygmy hog, and hispid hare are the wildlife of the zone. The diversity in plant communities and species is extremely high. The animal communities are also diverse. In fact, smaller carnivores exhibit a richness not seen anywhere else in the world. India's highest populations of elephants are here. The region also forms an important flyway on the route of migratory birds to and from Siberia and China. There are 17 protected areas covering 1,880 square km.

2.12.10 ZONE: ISLANDS

Andaman and Nicobar are a group of 348 islands that are biologically immensely rich. About 2,200 species of higher plants are found, of which 200 are strictly endemic. The avifauna comprises 225 distinctive species, of which 112 are endemic. This zone is India's richest in fish and coral communities. On these islands, there are 100 protected areas covering 708 square km. It is away from the nearest point on the main land mass and extend about 590 km. With a maximum width of 58 km, the island forests of Lakshadweep in the Arabian Sea have some of the best-preserved evergreen forests of India. Some of the islands are fringed with coral reefs. Many of them are covered with thick forests and some are highly dissected.

2.13 DISTRIBUTION OF FOREST IN INDIA

India is endowed with vast forest resources. Forests play a vital role in social, cultural, historical, economic, and industrial development of the country and in maintaining its ecological balance. They are the resource base for sustenance of its population and a storehouse of biodiversity. Other land use practices such as agriculture and animal husbandry are benefitted by forests. Realizing the crucial role of forests in maintaining the ecological balance and socio-economic development, the National Forest Policy 1988 aims at maintaining a minimum of 33% of country's geographical area under forest and tree cover. The forests in the country have been classified into 16 major types and 251 subtypes on the basis of climatic and edaphic features. Distribution of diverse forest types across the country is presented in Table 2.2.

2.14 BIODIVERSITY LOSS, BIOSPHERE INTEGRITY, AND TIPPING POINTS

Ecosystem management strategies aimed at maximizing conservation and public health co-benefits must consider that systems have emergent properties that are not possessed by their individual components: they are more than the sum of their parts. One example is the resilience of ecosystems to absorb shock in the face of disturbance (such as pests and disease, climate change, invasive species, or the harvesting of crops, animals or timber) and return to their original structure and functioning. Ecosystems can be transformed if a change in ecosystem structure crosses a given threshold. Structural changes may be manifested as a result of the removal of key predators or other species from the food web (Thomson et al., 2013), the simplification of vegetation or soil structure, increased or decreased aridity, species loss and many other factors. Biodiversity loss is continuing,

TABLE 2.2 Diversity and Distribution of Major Forest Types in India

Major Groups	Type and Group	Area (m ha)	Percent of Forest Area
Tropical Forests	Wet evergreen forest	4.5	5.8
	Semi-evergreen forest	1.9	2.5
	Moist deciduous forest	23.3	30.3
	Littoral and swamp forest	0.7	0.9
	Dry deciduous forest	29.4	38.2
	Thorn forest	5.2	6.7
	Dry evergreen forest	0.1	0.1
Sub-tropical Forests	Subtropical broad leaved	0.3	0.4
	hill forest	3.7	5.0
	Sub tropical pine forest	0.2	0.2
	Sub tropical dry evergreen forest		
Temperate Forests	Montane wet temperate	1.6	2.0
	forest	2.6	3.4
	Himalayan moist temperate forest	0.2	0.2
	Himalayan dry temperate forest	–	–
Sub-alpine & Alpine Forests	Sub-alpine forest	–	–
	Moist alpine scrub	3.3	4.3
	Alpine scrub	–	–

and in many cases, it is increasing (Butchart et al., 2010; Tittensor et al., 2014). Biodiversity loss has been identified as one of the most critical drivers of ecosystem change (Hooper et al., 2012). Changes in the diversity of species that alter ecosystem function may directly reduce access to ecosystem services such as food, water, and fuel, and alter the abundance of species that control critical ecosystem processes essential to the provision of those services (Chapin et al., 2000).

Ecosystem regime shifts, including "tipping points," have been widely described and characterized at local levels (e.g., eutrophication of freshwater or coastal areas due to excess nutrients; collapse of fisheries due to overfishing; shifts of coral reefs to algae dominated systems). There is a

ment, ecosystems, and corresponding services as well as to human health (McMichael and Beaglehole, 2000; Balmford and Bond, 2005; Cardinale et al., 2012; Myers et al., 2013). In many cases, the ecological implications are immense and the need to address them pressing if our planet is to provide clean water, food, energy, timber, medicines, shelter, and other benefits to an ever-increasing population. The rise in demographic pressures and consumption levels will translate into unprecedented demands on the planet's productive capacity, and concomitant pressures on the Earth's biological resources may undermine the ability of ecosystems to provide life-sustaining services (McMichael and Beaglehole, 2000; CBD, 2010).

The drivers (causes) of ill-health and human, animal, and plant diseases often overlap with the drivers of biodiversity loss. Some of the principal common drivers, identified in the third edition of Global Biodiversity Outlook (GBO 3) and reiterated in its fourth edition, include: habitat change, overexploitation, and destructive harvest, pollution, invasive alien species and climate change (CBD, 2010, 2014), all of which may be exacerbated by environmental changes.

The pressing need to jointly address both social and environmental determinants of health has been widely acknowledged through various multilateral agreements. However, the role of biodiversity as a mediating influence on human health (through the loss of ecosystem services, which are themselves mediated by ecological processes), while gaining more widespread attention since Rio, merits much more systematic assessment as well as more structured, coherent, and cross-cutting policies and strategies. These critical linkages should be translated into concrete policy targets as we embark on a new series of global commitments on sustainable development as the MDGs reach their term in 2015.

2.15.1 *HABITAT CHANGE*

Land-use change (e.g., full or partial clearing for agricultural production or natural resource extraction, such as for as timber, mining, and oil) is the leading driver of biodiversity loss in terrestrial ecosystems. Alteration of native habitats may also reduce resilience; for example, deforested areas may experience soil erosion, increasing ecological risks of extreme

growing concern that regime shifts could occur at very large spatial scales over the next several decades, as human–environment systems exceed limits because of powerful and widespread driving forces that often act in combination: climate change, overexploitation of natural resources, pollution, habitat destruction, and the introduction of invasive species (Barnosky et al., 2012; Leadley et al., 2014). Cardinale et al. (2012) suggest that the impacts of biodiversity loss on ecological processes might be sufficiently large to rival the impacts of climate change and many other global drivers of environmental change.

Leadley et al. describe scenarios for regional-scale shifts that would have large-scale and profound implications for human well-being (Leadley et al., 2014). The unprecedented pressures of human activity on biodiversity and on the Earth's ecosystems may also lead to potentially irreversible consequences at a planetary scale, and this prospect has led to the identification of processes and associated thresholds, and to the development of various approaches to define preconditions for human development on a planetary scale (Rockström et al., 2009; Barnosky et al., 2012; Steffen et al., 2015). Global efforts to pursue sustainable development will continue to be compromised if these critical pressures are not countered in a more rigorous, systematic, and integrated fashion.

2.15 DRIVERS OF CHANGE

In recent decades, the impact of human activity on the natural environment and its ecosystems has been so profound that it has given rise to the term anthropocene, popularized by Nobel prize-winning chemist Paul Crutzen, delimiting a shift into a new geological epoch, in which human activity has become the dominant force for environmental change (Crutzen, 2002). Anthropogenic pressures, demographic change, and resulting changes in production and consumption patterns are also among the factors that contribute to biodiversity loss, ill-health, and disease emergence. These pressures have shown a "great acceleration," especially in the past 50 years (Steffen et al., 2015). While some human-induced changes have garnered public health benefits, such as the provision of energy and increased food supply, in many other cases, they have been detrimental to the environ-

weather events such as sudden flooding, and limited food production potential from reduced soil enrichment. Furthermore, habitat changes such as deforestation directly alter the capacity of carbon sinks and thus further increase the risks of climate change.

Land-use change is also the leading driver of disease emergence in humans from wildlife (Jones et al., 2008). Changes to habitats, including through altered species composition (influenced by conditions that may more favorably support carriers of disease, as seen with malaria harboring vectors in cleared areas of the Amazon) and/or abundance in an ecosystem (and thus potential pathogen dispersion and prevalence), and the establishment of new opportunities for disease transmission in a given habitat, have major implications for health. Human-mediated changes to landscapes are accompanied by human encroachment into formerly pristine habitats, often also accompanied by the introduction of domestic animal species, enabling new types of interactions among species and thus novel pathogen transmission opportunities.

2.15.2 OVEREXPLOITATION AND DESTRUCTIVE HARVEST

Overexploitation of biodiversity and destructive harvesting practices reduce the abundance of the populations of species concerned, and in some cases, can threaten the survival of the species itself. Demand for wild-sourced food is increasing in some areas. The wildlife trade, for purposes such as supplying the pet trade, medicinal use, horticulture and luxury goods, is increasing globally, exacerbating pressures on wild populations. Practices for harvest, including unregulated administration of chemicals for the capture of animals (e.g., the release of cyanide or trawling practices for fishing) may also have impacts on non-target species, and/or unsustainable harvests may alter ecological dynamics, such as diminished potential for seed dispersion and implications for food chains (affecting also the humans who depend on them). As native biodiversity declines, local protein sources from subsistence hunting or gathering may be diminished, causing inadequate nutrition if alternatives are unavailable or lack necessary nutrients. Additionally, bushmeat hunting and consumption, sometimes in areas that have not been previously targeted for food sourcing (e.g., in newly established mining camps in formerly pristine habitat), may pose direct novel infectious disease transmission risks. Intensification of harvest and exploitative practices, such as the mixing of wildlife and domestic species in markets, as well as the mixing and spread of their pathogens, can create global epidemics, as seen with the 2003 outbreak of severe acute respiratory syndrome (SARS).

2.15.3 POLLUTION

Environmental pollution poses direct threats to both biodiversity and human health in many ways. Pollutant exposure risk is potentially increased for top-of-the-chain consumers such as humans and marine mammals through bioaccumulation along the food chain, as seen with mercury. Air pollution exposure presents risks of respiratory diseases. Other so-called "lifestyle diseases" (such as obesity and diabetes) may be influenced by access to physical fitness, which may be limited by outdoor and indoor air pollution levels. Chemicals such as pharmaceuticals or plastics containing endocrine-disrupting substances may be dispersed on entering water sources and other environmental settings, thus posing acute, chronic, or recurring exposures in humans and animals. Wide-scale application of antimicrobials for human and animal medicine and food production, much of which is excreted into the environment, is resulting in rapid changes to microbial composition, as well as driving development of antimicrobial resistant infections. Contaminated water may enable persistence of human infectious agents and their diseases, such as cholera-causing *Vibrio* and parasitic worm-transmitted schistosomiasis.

2.15.4 INVASIVE ALIEN SPECIES

Invasive alien species (IAS) pose direct threats to native and/or endemic species. The introduction of IAS may result in invasive species out-competing important food and traditional medicine sources for human populations, as well as causing fundamental impacts on ecosystems that may influence health processes. Examples of this include impaired water quality from the introduction of zebra mussel in the United Kingdom and North America, altered soil quality through the spread of weeds, and the reduced species decomposition facilitated by feral pigs grazing on native plants as well as agricultural land. In addition to these detrimental impacts, IASs pose risks of disease introduction and spread for native wildlife, agricultural species, and humans. As global trade and travel continues to increase, so do the health risks; changing climactic conditions may also enable establishment of IAS where climate would have previously limited survival, demonstrated with alarming clarity in the case of the pine mountain beetle invasion in Western Canada.

2.15.5 CLIMATE CHANGE

The direct and indirect impacts of climate change also pose risks for biodiversity and health; for example, shifts in species ranges may also facilitate changes in pathogen distribution and/or survival, as projected for Nipah virus (Daszak et al., 2013). Climate change also contributes to ocean acidification, coral bleaching, and diseases in marine life, as reef-building coral species are threatened with extinction. These in turn have significant implications for the large biological communities that coral reefs support and that sustain human health (Campbell et al., 2009). More extreme weather patterns and rising sea levels (*e.g.*, drought, flooding, early frost) may also be detrimental to food and water security, especially for population's dependent on subsistence farming and natural water sources. Human populations may also suffer acute health impacts from extreme weather (e.g., heat or cold exposure injuries).

2.15.6 DEMOGRAPHIC FACTORS INCLUDING MIGRATION

In addition to the direct drivers of biodiversity loss, large-scale societal and demographic changes, or intensified reliance on ecosystems for subsistence or livelihoods, often linked to biodiversity changes, may also impact vulnerability to disease. For example, new human inhabitants (recent immigrants) might not have immunity to zoonotic diseases endemic to the area, making them particularly susceptible to infection. Women who are required to butcher harvested wildlife, or men who hunt the game, may be particularly at risk. Moreover, those sectors of society that lack adequate income to purchase market alternatives may be more likely to access forest resources (including wildlife) for food and trade. Thus, there are likely socioeconomic and gender-specific relationships to these types of disease risks and exposures (WHO, 2008). Disease may also worsen the economic status of a population; vector-borne and parasitic diseases, the burden of which is driven by ecological conditions, have been shown to worsen the poverty cycle (Bonds et al., 2012).

2.15.7 URBANIZATION AS A CHALLENGE AND AN OPPORTUNITY TO MANAGE ECOSYSTEM SERVICES

Urbanization, the demographic transition from rural to urban, is associated with shifts from an agriculture-based economy to mass industry technology and service. With the majority of the world's population now living in urban areas and this proportion expected to increase, it is expected that urban health will become a major focus at the intersection of global public health and conservation policy. Urbanization is also closely linked with the social determinants of health, including development, poverty, and well-being. While urbanization is often associated with increasing prosperity and good health, urban populations also demonstrate some of the world's most prominent health disparities, in both low- and high-income countries. Rapid migration from rural areas as well as natural population growth are putting further pressure on limited resources in cities, and in particular, in low-income countries.

Much of the natural and migration growth in urban populations is among the poor. More than 1 billion people—one-third of urban dwellers—live in slum areas, which are often overcrowded and are affected by life-threatening conditions (UNDP, 2005). In low-income countries, disparities will continue to rise as the combination of migration, natural growth, and scarcity of resources makes it more difficult to provide the services needed by city dwellers (UN-Habitat, 2013). Poorly planned or unplanned urbanization patterns also have negative consequences for the health and safety of people, including decreased physical activity and unhealthy diets, which lead to increased risks for NCDs such as heart disease, cancer, diabetes, and chronic lung disease (WHO, 2010).

Urbanization also creates new challenges for biodiversity conservation; the development of cities is one of the most important drivers of land-use change. Moreover, it was estimated that up to 88% of protected areas likely to be affected by new urban growth are in countries of low-to-moderate income (McDonald et al., 2008). While cities typically develop in proximity to the most biologically diverse areas (Seto et al., 2011), relatively little attention has been paid to how cities can be more bio-diverse or to the importance of maintaining bio-diverse ecosystems for human health (Andersson et al., 2014). Several health benefits can potentially be derived from integrating biodiversity into urban planning schemes, and broader conservation and public health policies.

2.16 CONCLUSION

Population growth, and prevailing consumption and production patterns are part of these challenges; however, population movement, for example, also causes environmental stress. Conflict prevention can help reduce the environmental impact of large congregations of displaced persons in refugee camps (Burkle, 2010). Similarly, careful environmental impact assessment can mitigate the effect of communities displaced by large-scale development projects. In all these cases, a mixture of conservation, public health, social sciences, demography, and policy planning is needed to avoid the worst-case scenarios, wherein resource scarcity, ecosystem degradation, biodiversity loss, human conflict, and climate change combine to present the perfect storm of multiple, simultaneous, ongoing humanitarian crises. The architects of coherent strategic policies integrating the biodiversity and health nexus must harness these opportunities and reflect these imperatives for the health and well-being of present and future generations.

Conserving and sustainable management of biodiversity is critical to address the existing ecological crisis and sustain the future. However, in spite of the laws, we are rapidly losing our diversity that poses threat to our ecological balance and also contributes to climatic changes. We need to know where the lacuna is. Are not our laws, policies, and programs effective enough to deal with this ecological crisis? Or is it that the laws are not being implemented? Or people are not sensitized enough? I think all these factors are contributing to the existing conditions and rules and their implementation should be made more stringent.

KEYWORDS

- **biodiversity**
- **ecosystem**
- **forest**
- **geographical zones**
- **IPR**

REFERENCES

Andersson, E., Barthel, S., Borgstrom, S., Colding, J., Elmqvist, T., Folke, C., & Gren, A., (2014). Reconnecting cities to the biosphere: Stewardship of green infrastructure and urban ecosystem services. *Ambio.*, *43*(4), 445–453.

Balmford, A., & Bond, W., (2005). Trends in the state of nature and their implications for human well-being. *Ecology Letters*, *8*, 1218–1234.

Balvanera, P., Pfisterer, A. B., Buchmann, N., He, J. S., Nakashizuka, T., Raffaelli, D., & Schmid, B., (2006). Quantifying the evidence for biodiversity effects on ecosystem functioning and services. *Ecology Letters*, *9*(10), 1146–1156.

Balvanera, P., Siddique, I., Dee, L., Paquette, A., Isbell, F., Gonzalez, A., Byrnes, J., O'Connor, M. I., Hungate, B. A., & Griffin, J. N., (2014). Linking Biodiversity and Ecosystem Services: Current Uncertainties and the Necessary Next Steps. *BioScience*, *64*(1), 49–57.

Barnosky, A. D., Hadly, E. A., Bascompte, J., Berlow, E. L., Brown, J. H., Fortelius, M., Getz, W. M., Harte, J., Hastings, A., Marquet, P. A., Martinez, N. D., Mooers, A., Roopnarine, P., Vermeij, G., Williams, J. W., Gillespie, R., Kitzes, J., Marshall, C., Matzke, N., Mindell, D. P., Revilla, E., & Smith, A. B., (2012). Approaching a state shift in Earth/'s biosphere. *Nature*, *486*(7401), 52–58.

Bonds, M. H., Dobson, A. P., & Keenan, D. C., (2012). Disease ecology, biodiversity, and the latitudinal gradient in income. *PLoS Biology*, *10*(12), e1001456.

Butchart, S. H., Walpole, M., Collen, B., Van Strien, A., Scharlemann, J. P., Almond, R. E. A., Baillie, J. E. M., Bomhard, B., Brown, C., Bruno, J., Carpenter, K. E., Carr, G. M., Chanson, J., Chenery, A. M., Csirke, J., Davidson, N. C., Dentener, F., Foster, M., Galli, A., Galloway, J. N., Genovesi, P., Gregory, R. D., Hockings, M., Kapos, V., Lamarque, J. F., Leverington, F., Loh, J., McGeoch, M. A., McRae, L., Minasyan, A., Morcillo, M. H., Oldfield, T. E. E., Pauly, D., Quader, S., Revenga, C., Sauer, J. R., Skolnik, B., Spear, D., Stanwell-Smith, D., Stuart, S. N., Symes, A., Tierney, M., Tyrrell, T. D., Vié, J. C., & Watson, R., (2010). Global biodiversity: indicators of recent declines. *Science*, *328*(5982), 1164–1168.

Campbell, A., Kapos, V., Scharlemann, J. P. W., Bubb, P., Chenery, A., Coad, L., Dickson, B., Doswald, N., Khan, M. S. I., Kershaw, F., & Rashid, M., (2009). Review of the

Literature on the Links between biodiversity and climate change: Impacts, adaptation and mitigation. Secretariat of the convention on biological diversity, Montreal. *Technical Series No., 42*, pp. 124.

Cardinale, B. J., Duffy, J. E., Gonzalez, A., Hooper, D. U., Perrings, C., Venail, P., Narwani, A., Mace, G. M., Tilman, D., Wardle, D. A., Kinzig, A. P., Daily, G. C., Loreau, M., Grace, J. B., Larigauderie, A., Srivastava, D. S., & Naeem, S., (2012). Biodiversity loss and its impact on humanity. *Nature*, *486*(7401), 59–67.

CBD, (2006). *Agricultural Biodiversity*: A cross-cutting initiative on biodiversity for food and nutrition. https://www.cbd.int/decision/cop/default.shtml?id=11037 (accessed August 14, 2014).

Champion, H. G., & Seth, S. K., (1968). A revised survey of the forest types of India. Government of India Press.

Chapin III, F. S., Zavaleta, E. S., Eviner, V. T., Naylor, R. L., Vitousek, P. M., Reynolds, H. L., Hooper, D. U., Lavorel, S., Sala, O. E., Hobbie, S. E., Mack, M. C., & Díaz, S., (2000). Consequences of changing biodiversity. *Nature*, *405*(6783), 234–242.

Chauhan, B. S., (2012). Environmental studies. *Firewall Media*, *2008*, pp. 107–111. ISBN 978-81-318-0328-8. Retrieved 27 October 2012.

Chauhan, S. S., (2014). Status of biodiversity in India: issues and challenges. *Indian Journal of Plant Sciences*, *3*(3), 52–63.

Convention on Biological Diversity (CBD), (2010). *Global Biodiversity Outlook 3*. Secretariat of the Convention on Biological Diversity, Montreal.

Convention on Biological Diversity (CBD), (2014). *Global Biodiversity Outlook 4*. Secretariat of the Convention on Biological Diversity, Montreal.

Crutzen, P. J., (2002). Geology of mankind. *Nature*, *415*(6867), 23–23.

Daszak, P., Zambrana-Torrelio, C., Bogich, T. L., Fernandez, M., Epstein, J. H., Murray, K. A., & Hamilton, H., (2013). Interdisciplinary approaches to understanding disease emergence: The past, present, and future drivers of Nipah virus emergence. *Proceedings of the National Academy of Sciences*, *110*, 3681–3688.

Haines, A., Alleyne, G., Kickbusch, I., & Dora, C., (2012). From the Earth Summit to Rio+20: integration of health and sustainable development. *The Lancet*, *379*(9832), 2189–2197.

Hooper, D. U., Chapin III, F. S., Ewel, J. J., Hector, A., Inchausti, P., Lavorel, S., Lawton, J. H., Lodge, D. M., Loreau, M., Naeem, S., Schmid, B., Setala, H., Symstad, A. J., Vandermeer, J., & Wardle, D. A., (2005). Effects of biodiversity on ecosystem functioning: a consensus of current knowledge. *Ecological Monographs*, *75*, 3–35.

ICFRE, (2000). Forest Research and Education Report, Dehradun.

Jones, K. E., Patel, N. G., Levy, M. A., Storeygard, A., Balk, D., Gittleman, J. L., & Daszak, P., (2008). Global trends in emerging infectious diseases. *Nature*, *451*(7181), 990–993.

Kala, C. P., & Mathur, V. B., (2002). Patterns of plant species distribution in the Trans-Himalayan region of Ladakh, India. *Journal of Vegetation Science*, *13*, 751–754.

Kaul, R. N., (1970). Afforestation in arid zones. *Monographiae Biologicae*, *20*, The Hague.

Khan, T. I., & Frost, S., (2001). Floral biodiversity: a question of survival in the Indian Thar Desert. *Environmentalist*, *21*(3), 231–236.

Leadley, P. W., Krug, C. B., Alkemade, R., Pereira, H. M., Sumaila, U. R., Walpole, M., Marques, A., Newbold, T., Teh, L. S. L, Van Kolck, J., Bellard, C., Januchowski-

Hartley, S. R., & Mumby, P. J., (2014). Progress towards the Aichi biodiversity targets: An assessment of biodiversity trends, policy scenarios and key actions. Secretariat of the Convention on Biological Diversity, Montreal, Canada, *Technical Series*, *78*, pp. 500.
Leadley, P., Proença, V., Fernández-Manjarrés, J., Pereira, H. M., Alkemade, R., Biggs, R., Bruley, E., Cheung, W., Cooper, D., Figueiredo, J., Gilman, E., Guénette, S., Hurtt, G., Mbow, C., Oberdorff, T., Revenga, C., Scharlemann, J. P. W., Scholes, R., Smith, M. S., Sumaila, U. R., & Walpole, M., (2014). Interacting regional scale regime shifts for biodiversity and ecosystem services. *BioScience*, biu093.
Loreau, M., Naeem, S., Inchausti, P., Bengtsson, J., Grime, J. P., Hector, A., & Wardle, D. A., (2001). Biodiversity and ecosystem functioning: current knowledge and future challenges. *Science*, *294*(5543), 804–808.
Mace, G. M., Norris, K., & Fitter, A. H., (2012). Biodiversity and ecosystem services: a multilayered relationship. *Trends in Ecology and Evolution*, *27*(1), 19–26.
Maclaurin, J., & Sterelny, K., (2008). *What is Biodiversity?* University of Chicago Press.
McCann, K., (2007). Protecting biostructure, *Nature*, *446*(7131), 29–29.
McDonald, R. I., Kareiva, P., & Forman, R. T. T., (2008). The implications of current and future urbanization for global protected areas and biodiversity conservation. *Biological Conservation*, *141*, 1695–1703.
McMichael, A. J., & Beaglehole, R., (2000). The changing global context of public health. *The Lancet*, *356*(9228), 495–499.
Myers, S. S., Gaffikin, L., Golden, C. D., Ostfeld, R. S., Redford, K. H., Ricketts, T. H., Turner, W. R., & Osofsky, S. A., (2013). Human health impacts of ecosystem alteration. *Proceedings of the National Academy of Sciences*, *110*(47), 18753–18760.
Naeem, S., & Wright, J. P., (2003). Disentangling biodiversity effects on ecosystem functioning: deriving solutions to a seemingly insurmountable problem. *Ecology Letters*, *6*(6), 567–579.
Negi, S. S., (1996). *Biosphere Reserves in India: Land Use, Biodiversity and Conservation*. New Delhi: Indus Pub. Co., pp. 1–222.
Nelson, E., Mendoza, G., Regetz, J., Polasky, S., Tallis, H., Cameron, D., Chan, K. M. A., Daily, G. C., Goldstein, J., Kareiva, P. M., Lonsdorf, E., Naidoo, R., Ricketts, T. H., & Shaw, M., (2009). Modeling multiple ecosystem services, biodiversity conservation, commodity production, and tradeoffs at landscape scales. *Frontiers in Ecology and the Environment*, *7*(1), 4–11.
Reiss, J., Bridle, J. R., Montoya, J. M., & Woodward, G., (2009). Emerging horizons in biodiversity and ecosystem functioning research. *Trends in Ecology & Evolution*, *24*(9), 505–514.
Rockström, J., Steffen, W., Noone, K., Persson, A., Chapin III, F. S., Lambin, E. F., Lenton, T. M., Scheffer, M., Folke, C., Schellnhuber, H. J., Nykvist, B., De Wit, C. A., Hughes, T., Van der Leeuw, S., Rodhe, H., Sörlin, S., Snyder, P. K., Costanza, R., Svedin, U., Falkenmark, M., Karlberg, L., Corell, R. W., Fabry, V. J., Hansen, J., Walker, B., Liverman, D., Richardson, K., Crutzen, P., & Foley, J. A., (2009). A safe operating space for humanity. *Nature*, *461*(7263), 472–475.
Schwartz, M. W., Brigham, C. A., Hoeksema, J. D., Lyons, K. G., Mills, M. H., & Van Mantgem, P. J., (2000). Linking biodiversity to ecosystem function: implications for conservation ecology. *Oecologia*, *122*(3), 297–305.

Seto, K. C., Fragkias, M., Güneralp, B., & Reilly, M. K., (2011). A meta-analysis of global urban land expansion. *PloSone, 6*(8), e23777.

Steffen, W., Broadgate, W., Deutsch, L., Gaffney, O., & Ludwig, C., (2015). The trajectory of the Anthropocene: The Great Acceleration. *The Anthropocene Review*, 2053019614564785.

Thompson, I., Mackey, B., McNulty, S., & Mosseler, A., (2009). *Forest Resilience, Biodiversity, and Climate. Change*. A synthesis of the biodiversity /resilience /stability relationship in forest ecosystems. Secretariat of the Convention on Biological Diversity, Montreal. Technical Series no. 43, 67 pp.

Thomson, G. R., Penrith, M. L., Atkinson, M. W., Atkinson, S. J., Cassidy, D., & Osofsky, S. A., (2013). Balancing livestock production and wildlife conservation in and around Southern Africa's transfrontier conservation areas. *Transboundary and Emerging Diseases, 60*, 492–506.

Tilman, D., Knops, J., Wedin, D., Reich, P., Ritchie, M., & Siemann, E., (1997). The influence of functional diversity and composition on ecosystem processes. *Science, 277*(5330), 1300–1302.

Tittensor, D. P., Walpole, M., Hill, S. L., Boyce, D. G., Britten, G. L., Burgess, N. D., Butchart, S. H., Leadley, P. W., Regan, E. C., Alkemade, R., Baumung, R., Bellard, C., Bouwman, L., Bowles-Newark, N. J., Chenery, A. M., Cheung, W. W., Christensen, V., Cooper, H. D., Crowther, A. R., Dixon, M. J., Galli, A., Gaveau, V., Gregory, R. D., Gutierrez, N. L., Hirsch, T. L., Höft, R., Januchowski-Hartley, S. R., Karmann, M., Krug, C. B., Leverington, F. J., Loh, J., Lojenga, R. K., Malsch, K., Marques, A., Morgan, D. H., Mumby, P. J., Newbold, T., Noonan-Mooney, K., Pagad, S. N., Parks, B. C., Pereira, H. M., Robertson, T., Rondinini, C., Santini, L., Scharlemann, J. P., Schindler, S., Sumaila, U. R., Teh, L. S., Van Kolck, J., Visconti, P., & Ye, Y., (2014). A mid-term analysis of progress toward international biodiversity targets. *Science, 346*(6206), 241–244.

UN Millennium Project, (2005). *Investing in Development: A Practical Plan to Achieve the Millennium Development Goals*. New York, pp. 1–356.

United Nations Human Settlements Programme (UN-Habitat), (2013). State of the world's cities 2012/2013: *Prosperity of Cities*. Routledge.

WHO, (2008). Closing the gap in a generation: Health equity through action on the social determinants of health, Geneva.

WHO, (2010). Hidden cities: unmasking and overcoming health inequities in urban settings. *Technical Report*. Centre for Health Development, and United Nations Human Settlements Programme Un-habitat.

Wildlife Research Institute of India, (2009). *Biodiversity and its Conservation in India*, Wildlife Research Institute of India, Dehradun.

CHAPTER 3

APPLICATION OF IPR TO CONSERVE GLOBAL GENETIC RESOURCES FOR COMMONS

KUMARI RAJANI,[1] GANESH PATIL,[2] RAVI RANJAN KUMAR,[3] TUSHAR RANJAN,[3] SANTOSH KUMAR,[4] and ARUN KUMAR[1]

[1] *Department of Seed Science and Technology, Bihar Agricultural University, Sabour–813210, Bhagalpur, India, E-mail: rajani.dsstiari@gmail.com*

[2] *Vidya Pratisthan's College of Agriculture Biotechnology, Vidyanagari, Baramati–413133, India*

[3] *Department of Molecular Biology and Genetic Engineering, Bihar Agricultural University, Sabour–813210, India*

[4] *Department of Plant Pathology, Bihar Agricultural University, Sabour–813210, India*

CONTENTS

ABSTRACT

Commonly owned genetic resources, for example, herds of sheep, deer, fruits, medicinal plants, reservoirs, and forests rich in macro- and micro-flora, etc., in particular have benefited from the combination of technological progress in life sciences and the information sciences. International cooperation in global life science research is enhanced by the advancement of science, development of new methods for the identification, long-term conservation (e.g., freezing, construction of cDNA library, in situ hybridization) and shipping of genetic resources. In addition to that, the developments in information technology dramatically expanded the possibilities of distributed coordination as well as diminishing the search costs for locating genetic resources held in collections throughout the world. The commercial value of small genetic resources has put pressure on the sharing ethos that is at the basis of the exchange of resources within the commons. In this challenging situation, it becomes important to explore how the principles of commons-based production can also be applied to the specific case of global and regional genetic-resource governance. As biophysical entities, most genetic resources are widely dispersed, whether originally in nature or as a result of human domestication. Over the last three decades, national governments, patent owners, farmers, plant breeders, researchers, and a diverse array of non-governmental organizations (NGOs) have engaged in a vigorous debate over how to conserve and utilize the world's plant genetic diversity. In this article, the issue of intellectual property rights (IPRs) is discussed on global commons regime, which would allow researchers, breeders, and farmers free and unfettered access to all genetic resources. There is a need for appropriate organizational forms, legal arrangements, and social practices, which can help to better secure the global user community's need to address issues of common concern which is also discussed. The key issue is how to build upon these initiatives and to put the incipient global genetic-resource commons on a solid institutional basis that will enable commons-based production to co-exist, whenever effective, with market-based and state-based contributions to collective goods is discussed in detail in this article, which would be more useful for the readers.

3.1 INTRODUCTION

Natural resources are being used by common peoples in localized communities since historic times. These resources offer a sustainable alternative to both private ownership as well as state-based governance of resources (Ostrom, 1990). The term "commons" is generally refer to any resource that is shared by a group of people at the local, community, or global level (Hess and Ostrom, 2007). There are several well-known examples of natural resource like forest management in tribal regions of Madhya Pradesh and North-East parts of India. Exploitation and access to these natural resources are restricted to and regulated by local communities, even though the property regime governing the resource can vary from private to common or state ownership.

Commonly owned genetic resources are mainly the individual components of biodiversity, for example, herds of sheep, cow, goat, and deer, fruits, medicinal plants, reservoirs, and forests rich in macro- and microflora. Commonly owned genetic resources have benefited from the combination of technological progress in life sciences as well as information sciences (Parry, 2004; Dedeurwaerdere, 2011). The advancement of modern science, development of new methods for identification, long-term conservation (e.g., freezing, construction of cDNA library, in situ hybridization), and shipping of genetic resources enhanced interest and international cooperation in global life science research. In addition to that, the developments in information technology dramatically expanded the possibilities of distributed coordination, as well as diminishing the search costs for locating genetic resources held in collections throughout the world.

On the other hand, the positive impact of the technological changes on the development of global genetic-resource commons has been attenuated by a set of equalizing factors, which could expose the whole enterprise (Reichman et al., 2009; Dedeurwaerdere, 2011). The commercial value of a small subset of genetic resources, such as in the field of say for example food article (Basmati rice and Turmeric) or pharmaceutical product development, has put pressure on the sharing culture that is at the basis of the exchange of resources within the commons. In general, laws against secrecy have been downplayed by delays in publication and restrictions on the sharing of research materials and tools, which in turn have often been caused by concerns about intellectual property rights (IPR) (Rai, 1999). Another constraint is the heterogeneity of legal frameworks and institutional rules that operate at the global scale. The most important hurdle in this regard is divergent national access and benefit sharing legislation across countries, and a lack of international coordination in the implementation of legal provisions in a way that is consistent with the needs of public science in developing and industrialized countries (Roa-Rodríguez and van Dooren, 2008; Jinnah and Jungcurt, 2009).

The policymakers and genetic resource managers have mainly focused on inventing new methods for organizing and integrating diverse collections of resources, with a view to better securing the various user communities' research needs (Dedeurwaerdere, 2011). In a report of the Ad Hoc Open-Ended Working Group on Access and Benefit-Sharing (2005), it was recommended that an international regime on access to genetic resources and benefit-sharing directed toward conservation, health, agriculture, and related goals needs to be developed in a more equitable way than the patent system and that alternative *sui generis* models such as "open source" style models should be developed (Jeffery, 2008).

The need for such alternative models has already been acknowledged in the field of genetic resources used in food and agriculture, where it became clear and accepted. In contrast to the case of medicinal plants, a purely bilateral system would not be possible, because the plant breeding process calls for a broad range of genetic resources as inputs into any one product (Burhenne-Guilmin, 2008). It showed that, more generally, when the research and innovation process is based on screening or off breeding from pools with multiple inputs from various sources, commons-based innovation offers an interesting institutional option, as an alternative to both private proprietary and state-based solutions. Commons-based advancement in genetic resources allows both the obstacles of contracting over every single entity in a system of exclusive property rights (Dedeurwaerdere, 2005), and the lack of flexibility of centralized governmental and inter-governmental organizations (Halewood, 2010) to be overcome, without interfering downstream commercial applications (Benkler, 2006; Reichman et al., 2009). In this challenging situation, it becomes important to explore how the principles of commons-based production can also be applied to the specific case of global and regional genetic-resource governance.

3.2 PUBLIC GOODS AND PUBLIC SCIENCE

Global public goods (GPGs) are the socio-economic terminology for a vast range of goods and services that welfare everyone, including stable climate, public health, etc. Therefore, economic security and sustainment of GPGs is a considerable governance challenges (Dedeurwaerdere, 2011). At the national level, public goods are often provided by government, but at the global level, there is no established state-like entity to take charge of their provision. The intricate nature of many GPGs poses additional problem of coordination, knowledge generation, and the formation of citizen preferences.

Global pools of biomaterials are needed in the context of the challenges of food security, global health issues and the biodiversity crisis more generally (Dedeurwaerdere, 2011). Similarly, the genomics revolution and the broader impact of the globalization of research in the modern sciences have enhanced the interest and cooperation in the collection of genetic resources. As a result, huge number of human, animal, plant, and microbial genetic materials are collected throughout the world from various regions and habitats and exchanged in collaborative research networks.

During 1980s, Africa faced the destruction of a major crop, Cassava, by a scale insect, the mealy bug (Hammond and Neuenschwander, 1990). The research work was conducted in Latin America on the natural enemies of this bug, which resulted in the identification of a predator that was later imported into Africa and successfully used in a major biological control program in Cassava. As a result, millions of dollars of food crops were saved. The sharing of biological resources like microbial materials (e.g., nitrogen-fixing microorganisms, enzyme-producing microbes, etc.) is the classical example of worldwide sharing of biological resources Based on the worldwide exchange of some well-characterized and high performing isolates of these bacteria, they are being used in public and private research, for training and education, and commercially produced in large quantities in various countries (Dedeurwaerdere et al., 2009).

The increase in the exchange of genetic materials has raised a set of new collective-action problems. One of the main problems is the increase in practices that potentially create new threats to food and agriculture, quality management (Stern, 2004) and to human health (Doyle et al., 2005). These problems has led to initiatives to promote the exchange networks in truly globally distributed pools with common quality standards, clear rules for entry into the pool, and coordinated management like creation of a prototype for the Global Biological Resources Centers Network (GBRCN) in the microbial field (Smith, 2007) and/or the new proposed coordination structure of the Collaborative Group on International Agricultural Research in the field of crop genetic resources (CGIAR, 2003, 2009).

Digital infrastructures may create a new set of methods for breaking down and rebuilding the group business/project. The use of computer-based tools and ways of doing things in life sciences makes it possible to build large knowledge storage places, and to develop data mining tools for huge collection of data in the distributed network of storage places into a virtual collection (Dawyndt et al., 2006).

Digital networks also make it possible to directly improve the worldwide exchange of materials, by spreading around and coming together around common machine-readable Material Transfer Agreements (Nguyen, 2007). Finally, in an organized way recording/writing down the source and history of the materials deposited in genetic resources collections, and releasing this information online, the digital information has also become a tool for making the back-and-forth exchanges clearly visible (Fowler et al., 2001).

At present, genetic resources collections are taking advantage of the development of these new methods, by networking the existing infrastructure of physical collections into worldwide digital data and information (Dedeurwaerdere, 2011). The aim of this chapter is to discuss the valuable contribution of these new models and methods from the digital information commons to the further institutionalization of the exchange networks into truly around the world distributed pools.

3.3 MODELS FOR DESIGNING BIOLOGICAL RESOURCE COMMONS ON A GLOBAL SCALE

The design of global genetic resource commons ecosystem should take into account the important features of useful components of genetic resources. Genetic resources are complex products, with both biophysical thing and informational part (Dedeurwaerdere et al., 2011). As biophysical things/businesses, most genetic resources are widely dispersed, whether novel in nature (Beattie et al., 2003), or as a result of human domestication (Braudel, 1992). As a result, it is often expensive to leave out users from assessing the natural resources.

In several cases, biological components are accessed not only for direct exploitation of the entities itself, but also for access to the informational components (Goeschl and Swanson, 2002b; Dedeurwaerdere, 2005). For example, large quantities of biological articles are collected in order to screen the biological functions and properties they exhibit against certain targets. Once a new object or function is discovered, genetic similarity searching can identify the genetic sequences that are involved in the expression of these properties (Dedeurwaerdere et al., 2011). This may in turn lead to further research on these genes, without having to access the particular organism that lead to the discovery of the new informational inputs. Nevertheless, using particular things becomes important at the end of the research and invention of new things chain, when biological things are developed for commercial uses. Therefore, any government in power for controlling access to these valuable supplies should take into account both the broad informational features of the pool of useful entities and the possible commercial uses of specific biological entities.

Genetic resources are informational inputs in the process of research and innovation, both as stocks (accumulating important traits) and as generators of new flows of information (Swanson and Goeschl, 1998). The global genetic resources are presently regulated by their uniqueness. One set of regulations, personified in the access, and profit-sharing system established through the Convention on Biological Diversity, focuses on genetic resources as material goods (CBD, 2002). Most of the debates around these regulations have been elicited by the need to regulate those natural resources that are exchanged for their known or likely commercial value (Safrin, 2004; Reichman et al., 2009). Another set of regulations addresses the informational components, but mainly with the aim of creating incentives for private venture in these resources at the end of the innovation chain (Dedeurwaerdere et al., 2007; Jeffery, 2008). In both cases, the specific characteristics of research based on the screening and analysis of the informational components of large pools of resources is not considered.

It might be more prolific to look at the institutional solutions and models developed in the related field of the digitally networked information commons in order to take the specific informational features of the networked genetic resources into account (Dedeurwaerdere, 2011). Digital information commons have been proven to offer a set of vigorous and successful models for the production of informational goods and services (Lessig, 2001; Hess and Ostrom, 2007; Boyle, 2008). Moreover, there is already considerable experience with these global commons, and systematic research on basic design principles has been conducted. This can provide elements for a systematic analysis with the genetic resource commons.

This section focuses on two key common design principles of successful commons that came out of this research, which are the role of non-market motivations and the modular character of the organizational architecture. The main organizational feature that is common to all successful digital information commons is the design of complex incentive schemes that are driven more by social and fundamental motivations than by financial rewards (Benkler, 2006). Diverse motivations are common in a heterogeneous set of initiatives such as open source software communities, global genetic sequence databases, and distributed peer-to-peer computational infrastructures. Because of the trouble of putting a precise monetary value on the creative inputs of a vast and distributed network of contributors, it has proven to be more effective to rely on non-market motivations for organizing the networks (Deek and McHugh, 2008).

In addition, extensive practical research has shown that when social motivations are involved, such as increasing appreciation in a collaborative group or the satisfaction of intrinsic motivation with respect to furthering general interest objectives, financial rewards can decrease the willingness to contribute to the global pool (Frey and Jegen, 2001). Further, there are hidden costs to the move from social to financial rewards. These include the costs related to a clear explanation of the tasks to be paid for (Deci, 1976) and an economic evaluation of the value of each and every input to these tasks (Benkler, 2006).

The exchange of genetic resources in the global commons is clearly a case where social and essential motivations will play a significant role. Indeed, the attribution of a monetary value to each entity is especially hard, or simply impossible, when the genetic resources used as inputs for collaborative research in global exchange networks have to be assessed (Dedeurwaerdere, 2011). Many innovations result from the combination and comparison of information gained from a wide variety of genetic resources from different sources, which all play a certain, varying, role in the progress of the research. In some cases, the initial value of the resource may be increased by informational inputs that are difficult to quantify, such as associated know-how and traditional knowledge, but it can make a major contribution to research into environmental, food, and health-related properties (Blakeney, 2001).

The second attribute, which plays a key role in the success of commons-based production of knowledge in the digital commons, has been the adoption of modular scientific and organizational architectures. Modular architectures have allowed efforts and assistance from many human beings, which are diverse in their quality, quantity, focus, timing, and geographical location, to be pooled in an effective manner (Benkler, 2006). Modularity presumes the presence of a set of independently produced components that can be integrated into a whole. The fine-grained character of the modules determines the number of potential contributors to the network. If there is a large set of relatively small contributors, each of whom only has to invest a moderate amount of additional effort and time in the network, the potential benefits of taking part in global exchange networks is likely to be high. However, if even the smallest contributors are relatively large, and if they each require a large investment of additional time and effort to take part in the collaborative network, the potential reciprocity benefits of being part of the network, and the cost-effectiveness of doing so, will diminish and the universe of potential willing contributors will probably decrease. Modularity is undoubtedly also present in major successful collaborative projects in the field of the genetic resource commons, such as the collaborative sequencing of the 1,000 genome sequencing project undertaken at the University of Alberta (Goff et al., 2011; Matasci et al., 2014).

The importance of nonmarket motivation is a necessary circumstance for the emergence of effective commons-based production, but it is clearly not sufficient (Dedeurwaerdere, 2011). It is the combination of the potential of nonmarket production of collective goods, and the success of an organizational form that allows wide contributions to be incorporated, that makes effective commons-based innovation possible on a global extent. Research on these general design principles shows that under conditions of appropriate quality control, and an initial investment in the creation of social networks (Benkler, 2006), commons-based production and management of informational goods can be a popular and valuable institutional modality that co-exists with market or state-based production of knowledge goods. This is especially true in the early stages of research on the novelty and product-development chain, when access to multiple inputs is required (Dedeurwaerdere, 2011).

3.4 GENETIC RESOURCES FOR FOOD AND AGRICULTURE

The impact of the intellectual property regime on access to genetic resources is much greater than in the plant genetic-resources as compared to the microbial or animal genetic resources field (Chen and Liao, 2004; Tvedt et al., 2007). Plants are well-defined varieties and have greater genetic stability on reproduction than microbes or animals. It shows that exclusive rights can extend to direct offspring and the results of all cross-breeding from this offspring that has sufficient genetic similarity among protected cultivars or contains a specific gene. On the one hand, the ease of transfer of traits between crops through different breeding methods makes it very difficult to protect the proprietary information contained in improved varieties or to stimulate private investment in the absence of IPR (Swanson and Goeschl, 1998). On the other hand, IPR favor the innovators who are already situated on the innovation frontier (Goeschl and Swanson, 2005), under-represent the needs of poor countries (Benkler, 2006), and do not provide appropriate incentives for collaborative investment in the long-term informational values associated with the resource (Dedeurwaerdere et al., 2007).

The global crop commons look out to address these and other problems that exclusive rights regimes have created for innovation, especially in the experimental breeding sector (based on the systematic cross-breeding in crops) (Dedeurwaerdere, 2011). The global crop commons is based on assisted access to materials and their derivatives within a common pool of the world's major crops and forage plants for research and breeding purposes (Byerlee, 2010). Initially, this pool was built in the context of the international experimental breeding programs organized by the Collaborative Group on International Agricultural Research (CGIAR). This regime was further formalized, in response to the threats posed by the intellectual property regime, through the adoption of the International Treaty on Plant Genetic Resources of the Food and Agriculture Organization in 2001.

Extensive research is conducted on the institutional arrangements adopted within the global crop commons (Helfer, 2005; Halewood and Nnadozie, 2008). In 2011, the crop commons, formalized through the International Treaty, pools over 1.2 million accessions conserved in the collections and genebanks of contracting parties all over the world (Dedeurwaerdere, 2011). The majority comes from the 11 international collections of the CGIAR, few from other international collections, while more and more national public collections are officially joining the multilateral system of exchange as the treaty is executed (www.planttreaty.org/inclus_en.htm). The plant genetic resources that are within this pool are exchanged with the viral license of the treaty, for uses of these materials for research, breeding and education purposes. When commercial applications are developed, the Treaty offers (1) commercialization with a non-exclusive use license, which permits further use for non-commercial research, breeding and education purposes or (2) commercialization with an exclusive-use license and the payment of a fixed royalty to a multilateral fund.

Nonmarket values and modular organization also play a prime role in making the crop commons a sustainable institutional form. Collaborative research of scientists from different countries for crop-improvement programs underlie the exchange practices and promote the sharing of information, and the integration of regional efforts. The dedication toward the common goal of increasing food production, and global poverty reduction, is a key driver of the whole system (Byerlee, 2010). During early days of the crop improvement program, the community also invested in strengthening these social norms. In the field of wheat improvement alone, about 1400 individuals from 90 countries have participated in these training courses, and more than 2000 scientists have visited the International Maize and Wheat Improvement Center (CIMMYT) in Mexico. In addition, participating countries are allowed to give their own names to the varieties they release. This produces a sense of ownership and ensures that the international seed banks are seen as honest brokers with respect to germplasm and information sharing (Byerlee, 2010). Finally, the CGIAR has developed policy guidelines that broadly reflect these values, both before (CGIAR, 2003) and after (CGIAR, 2009) the ratification of the International Treaty.

Experimental breeding is a better example of a worldwide modular and distributed organization of research and innovation. One well-documented case is the international nursery network organized by CIMMYT. Every year, CIMMYT dispatches improved germplasm to a global network of wheat research cooperators who evaluate wheat germplasm in experimental trials targeted at specific agro-climatic environments (Dedeurwaerdere, 2011). Between 1994 and 2000, CIMMYT distributed 1.2 million samples to over 100 countries, corresponding to an average of 500 to 2000 globally distributed field trials per year (Byerlee, 2010). Data from the field trial were submitted back to CIMMYT for analysis, and the results were shared to the network of collaborating scientists. In this way, the crop commons builds an iterative collaborative platform that collects environmental and local feedback in a similar way to that in which bug reports are collected by free software projects (Benkler, 2006).

Many similarities have been observed between the microbial commons and the global crop commons after the analysis of these institutional characteristics. In both fields, institutional arrangements have established a globally networked commons, which is open to new users and contributors to the system under standardized non-exclusive contracts. Based on the quantitative data, it is observed that the scope of the crop commons is more limited as compare to the microbial commons, which covers far more individual collections and has a larger number of holdings. Nevertheless, within the crop commons, all the material is exchanged under a formal viral license, because of the major threat of exclusion from key research resources through patents, in the microbial commons. Depending on the circumstances and the commercial pressure on the collections, a mix of formal and informal contracts is used.

3.5 FARM ANIMAL RESOURCES

For the management and exchange of genetic resources in animal breeding, three major institutional arrangements are being used. First, the development of hybrid breeding sector which is based on crosses of very different parent or grandparent lines. As innovators do not disclose the parent and grandparent lines that are used in the production of hybrid, unauthorized reproduction of animals can be effectively prevented through technological means. As a result, in areas where hybrid breeding is a well-developed technological option—mainly poultry and pigs—an exclusive access regime has developed within a centralized and large-scale breeding industry (CGRFA, 2009).

Most livestock breeding is based on experimental breeding within a pool of animals that are managed in an open commons. On the one hand, when animals are exchanged between livestock keepers, the assumption is normally that the owners of the breeding animals acquired through such exchanges can use the genetic resources involved for further breeding (CGRFA, 2009). On the other hand, sustainable breeding requires a high level of coordination; therefore, the majority of experimental breeding programs that are run by farmer-owned co-operatives and breeder organizations operate in the context of national breeding programs or farmer-driven societies with a regional scope (CGRFA, 2009). Such programs are developed by one country/one region, even if there is often an important level of cross-breeding with imported animals to improve the genetic quality of the pool. As such, this institutional arrangement for commons-based management is not a globally interconnected pool as in the case of plants and microbes, but can be better characterized as a global network of exchange among limited (national or regional) commons.

Commons-based experimental breeding is, however, increasingly coming under demands from international companies that are taking over farmer-owned cooperative schemes, especially cattle-breeding schemes (Mäki-Tanila et al., 2008). This has led to the development of a third institutional regime, based on the operations of centralized commercial-breeding companies, with high expectations of quick earnings and a unilateral focus on productive characteristics. This centralization of breeding operations may raise new challenges, such as the reported decline in the reproduction and health traits of the Holstein breed, one of the most widely used dairy cows, due to a long-standing importance on production yield (Mäki-Tanila et al., 2008). Other challenges posed by the global commercial breeding companies are the introduction of new business practices like patents. However, these patents mostly concern certain genetic mutations causing genetic defects, while patents on productive traits at present only have a minor impact because of the multi-locus nature of most economically important traits (Mäki-Tanila et al., 2008).

In the organization of the traditional commons-based production sector in animal breeding, nonmarket values play an important role, described above, although these values have to be combined with the productivity constraints of the private farms that are breeding the animals (Mäki-Tanila et al., 2008; CGRFA, 2009). Animal breeding is part of national food security, and cooperative breeding programs are set up to promote collective goals such as animal health and the conservation of genetic variety within populations and breeds (which is essential to meet future challenges in the development of livestock) (Dedeurwaerdere, 2011). These nonmarket motives have to be shared by most of the members of the collective pool to be effective. This explains why most schemes are developing common guidelines for quality management and sustainable breeding. They are also actively promoting these guidelines among individual farmers, through information campaigns and quality-assurance contracts provided by the breeders' cooperatives. Finally, in many countries, legal rules have been adopted to strengthen the general interest objectives of the breeders' organizations (FAO, 2007). For instance, under current regulations, existing breeding organizations cannot claim property rights on the basis of which they could breed the animal in question exclusively. Moreover, any new breeding organization has to be state approved, comply with a set of quality management standards, and undertake conservation-breeding programs.

In order to deal with the specific problems of animal breeding, the modular organization of the experimental breeding program has been developed as a solution, such as the need to limit inbreeding and to maintain a sufficiently diverse breeding base for disease management (Dedeurwaerdere, 2011). The goat improvement program is developed in France by Capgènes, which exemplifies this modular organization (www.capgenes.com). In this program, 1,000 best-performing animals are selected annually from a pool of 170,000 goats on 800 farms. From these 1,000 animals, 40 male goats are selected after a lengthy process of quality checking and off-breeding. These males then serve as the starting point for the following year's artificial insemination program for breed improvement.

As can be seen from this analysis, there are some major differences between the institutional characteristics of commons-based production with animal genetic resources and those using plant and microbial resources. The main differences are the reliance on private actors (rather than public collections) whose resources are pooled in a collective breeding program, and the limited geographic scope of the commons-based improvement programs (Dedeurwaerdere, 2011). The latter remains true in spite of the active international exchange of genetic material for the selective upgrading of domestic breeds, which creates a network of highly inter-related populations in various countries. The greatest institutional similarities are observed between the animal and the microbial sector. In these two sectors, many exchanges still happen on an informal basis, because the threat of possible misappropriation through patents or breeders' rights is relatively weak (Dedeurwaerdere, 2011). However, the recent introduction of new business practices may lead to a rapid change in this situation.

3.6 AGRICULTURAL RESEARCH AND INTELLECTUAL PROPERTY RIGHTS

The adoption of a set of binding rules to govern transactions involving living resources would alleviate many problems caused due to the lack of standardization and agreed collective intellectual property rules, which would preserve the salient features of the current system of exchange (Dedeurwaerdere, 2011). To the extent that an efficacious standard material transfer agreement among all community members (livestock breeders' network) or all parties to a consortium (for pooling a set of common resources) would harmonize the current practices, it would set the basis for a fact-based commons for the exchange of biological material. The primary questions that need answer is that the extent to which such a commons would be truly community/livelihoods friendly, and the further extent to which such a community/livelihoods friendly regime could implement the potential for biodiversity-based innovation inherent in the new opportunities for food and agriculture production from large-scale breeding and screening of useful living material in the live sciences (Hall, 2004; Kurtböke, 2004).

Important steps in that direction could be made by the systematic adoption in all public culture collections of measures that include:

- Introduction of standard material transfer agreements (MTAs) for the distribution of materials among all community/pool members, thereby preventing a race to the bottom, by either providers (who might impose more restrictions) or users (who might block access to innovations based on secrecy or exclusive intellectual property). Such a framework agreement could build upon elements from the standard MTA adopted for plant genetic resources in the International Treaty for Plant Genetic Resources for Food and Agriculture;
- Allowing collections of biological materials to redistribute materials to collaborating scientists and farmers, or to other collections that operate under the same framework agreement (e.g., the case in the ECCO MTA discussed above).

In short, there is a "bundle of rights" attached to biological resources that should be regulated by laws and managed through agreement and contracts. The question of "who owns what" may be modified to "who has what rights" over microbiological resources.

Communities, farmers, and biological resource managers clearly have rights in the material they manage and preserve. Having these rights, they also can specify the conditions under which they distribute their material. So, ownership allows the community to define when biological material is shared on a non-exclusive basis and, conversely, when a restrictive licensing policy is justified.

In this way, the concepts of the "genetic resource commons" on the one hand, and "intellectual property rights" (exclusive use), on the other hand, are not opposed, but become two harmonizing tools at the disposition of the culture collection community (Chen and Liao, 2004; Dedeurwaerdere, 2007).

In the field of the genetic resource commons, there are some budding examples of models of collective intellectual property management leading to the building of a community/livelihoods-based genetic resource commons. As stated above, one prominent example is the model of the 2001 International Treaty for Plant Genetic Resources for Food and Agriculture.

3.7 THE INTERNATIONAL TREATY ON PLANT GENETIC RESOURCES

The International Treaty on Plant Genetic Resources for Food and Agriculture (ITPGRFA) covers an integrated commons linking both material and digital data and databases within a distributed and open access framework. It has a larger framework in its federated network of genebanks and of information portals that is now being established, which is being administered by a governing body made up of the 121 contracting parties of the treaty. The international treaty generates a multilateral system of access and benefit-sharing and implements the access and benefit-sharing principles on a multilateral basis. As we are aware, it takes about 1 or 2 years to negotiate a well-done bilateral access agreement, and they may be negotiated on a case-by-case basis. In contrast, under the ITPGRFA multilateral system, more than 600 transfers have been made every day of plant genetic material related to agriculture. It may be impossible to negotiate access and benefit-sharing on a bilateral case-by-case basis for all these transfers. Instead, the multilateral system laid out by the international treaty provides a low transaction cost, pooled commons of genetic material. Other components of the international governance architecture include the 1961 International Union for the Protection of New Varieties of Plants (UPOV); the 1995 Agreement on Trade-Related Aspects of

Intellectual Property Rights (TRIPS), and the various intellectual property agreements that are incorporated by reference under the TRIPS standards, such as the Paris Convention; and the 1985 Budapest Treaty on Deposit of Microorganisms for Purposes of Patent Procedure. The Consultative Group on International Agricultural Research (CGIAR) is not a legal body, but rather the largest network of agricultural research centers and genebanks of plant genetic resources in agriculture, and it plays an important role in the treaty (Chen and Liao, 2004; Dedeurwaerdere, 2007).

The timeline leading up to the ITPGRFA can be traced back to the CBD, which was adopted in 1992. In 1994, a request was transmitted to the Food and Agricultural Organization (FAO) of the United Nations to revise a pre-existing soft law instrument into a binding framework that would be in harmony with the access and benefit-sharing principles of the CBD. The negotiations lasted for 7.5 years, and finally, in 2001, the treaty was adopted by the FAO Conference, coming into force in 2004. The first governing body session under the treaty was held in 2006, and within 3 years following that session, 40 instruments of ratification were deposited. In 2006, 2 years after the ITPGFRA came into effect, the first session of the governing body adopted the Standard Material Transfer Agreement (SMTA). It was the fastest rate of authorization of an FAO-administered treaty in the historical prospect of FAO. The SMTA is the standard contract for transferring genetic material within the multilateral system. The third session of the governing body was held in June 2009 in Tunisia, and the next session held in March 2011 in Bali. This treaty creates a multilaterally governed gene pool. Genetic material is put into that gene pool by various actors. These include the nations those ratified the treaty and also international organizations, such as the International Atomic Energy Agency, which included its mutant germplasm repository in this gene pool and is now also being governed by the terms of the treaty. Individuals and organizations also contribute genetic material, including private sector entities and indigenous and local communities, such as the Quechua communities from Peru, who deposited their own germplasm into this gene pool. There are large numbers of stakeholders who are depositing material. A major contributor is the group of international agricultural research centers of the CGIAR, which have so far deposited the bulk of material with solid documentation. Once material is included in the gene pool, its use is governed by a chain of SMTAs. It begins with the provider of the material, transferring it under a standard material transfer agreement, call it SMTA1, to the recipient. This recipient can become a provider and transfer the material under a second agreement, SMTA2, to a second recipient. By receiving the material under the SMTA, the recipient takes on an obligation to transfer this material to other parties only under the same terms and conditions specified in the original SMTA. This builds a contractual chain that eventually leads—after a long series of transfers that, in plant breeding, often takes 5 to 10 years—to the development of a commercial product. The mechanisms and operational systems of the ITPGFRA are illustrated in Figure 3.1.

3.8 USE OF INTELLECTUAL PROPERTY RIGHTS (IPR) TO PRESERVE THE GLOBAL GENETIC COMMONS

Over the last 30 years, national governments, patent owners, farmers, plant breeders, researchers, and a diverse array of nongovernmental organizations (NGOs) have engaged in a vigorous debate over how to conserve and utilize the world's plant genetic diversity. One group advocates a pure

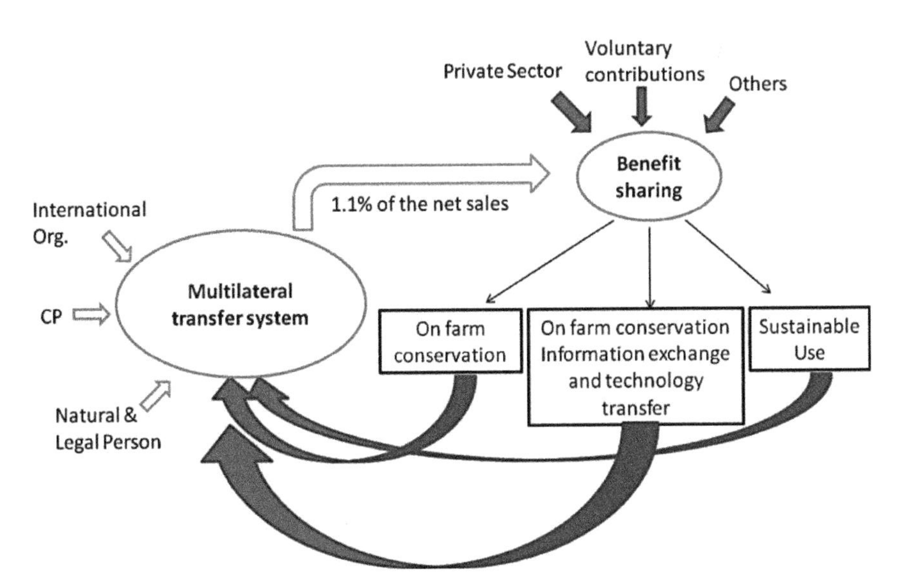

FIGURE 3.1 The main operational systems and mechanisms of the International Treaty on Plant Genetic Resources for Food and Agriculture.

global commons regime, which would allow researchers, breeders, and farmers free and unregulated access to all plant genetic resources (PGRs), including those held in international seed banks, in national collections, and in situ on public lands. On the other side, other group advocates for private property approach, which seeks to encourage plant-related innovations in agriculture and biotechnology by allowing isolated and modified genetic resources to be owned by patentees and commercial plant breeders. A critical issue in this ongoing controversy concerns the role of IPRs, and, in particular, where to draw a boundary between plant genetic materials that must remain in the public domain and those that can be privatized. Debates were to fix this boundary date back to the "seed wars" of the early 1980s, during which governments in developing countries pressured the Commission on Genetic Resources for Food and Agriculture (CGRFA) to stop the flow of PGRs from centers of biodiversity in the developing world to plant breeding industries in industrialized nations. These governments argued that commercial plant breeders were using PGRs to develop new proprietary plant varieties without compensating the countries that provided the raw materials for their innovations.

If that product is not available without restrictions for further research, training, and breeding—for example, if the product is under a patent claim—then the SMTA specifies that the recipient shall pay 1.1% minus 30% of net sales of that product to the beneficiary fund of the ITPGRFA. This is a multilaterally created, governed, and administered trust fund that receives proceeds from products that incorporate material received from the gene pool. This beneficiary fund also receives funding from a series of other channels, such as voluntary contributions from contracting parties. Several governments, including Norway, Spain, Italy, and Switzerland, have given voluntary contributions. At the opening of the Svalbard Global Seed Vault in Norway, Prime Minister Stoltenberg along with the Minister of Agriculture announced that Norway will each year contribute 0.1% of all national seed sales to the beneficiary fund of the treaty. Contributions also come from the private sector, including humanitarian grant-making institutions, foundations, and individual donor. The accumulated funds are dispersed according to multilaterally agreed-upon funding priorities, selection criteria, and operational procedures. For example, the treaty specifies a priority for funding to farmers in developing countries who conserve and sustainably use plant genetic diversity. At the second session of the governing body, three funding priorities were set by the governing body: on-farm conservation of plant genetic diversity, information exchange and technology transfer and sustainable use of plant genetic resources, including through characterization, research, and participatory plant breeding. These funding priorities are intended to further conservation and maintenance of genetic diversity, which in turn feeds the global gene pool, which is established by the treaty. In short, the treaty is intended to create a virtuous circle, overseen by the governing body of the treaty. That governing body now includes 121 governments, with additional ratifications being underway at present. The system started up in 2007, and since that time, there have been more than 1.2 million accessions of plant genetic material that we know of. That last qualifier is important because in reality, there have certainly been far more accessions, but we do not know of all the material that is included in the system. Getting complete and reliable datasets specifying which material is included in the system is itself a major undertaking that is now under way. In the last 12 months, there were about 440,000 transfers by the CGIAR alone. This does not include transfers of material from regional and national gene banks. There were more than 600 transfers each day worldwide. On the benefit-sharing side, in June 2011, the benefit-sharing fund disbursed half a million dollars in grants for the conservation of crop genetic diversity and the sustainable use of genetic resources. The governing body has adopted an objective of raising $160 million over the next 5 years for the benefit-sharing fund, with a planning target of $50 million. It also adopted a strategic plan, which lays out the mobilization of these resources over the next 5 years (Bhatti, 2011).

Pertaining to the exchange of information, the treaty states that access to plant genetic resources that are protected by intellectual and other property rights shall be consistent with the relevant international agreements and minimum standards for the availability, exercise, and enforcement of IPRs, and shall also be consistent with relevant national laws. The treaty also states, however, that recipients shall not claim any IPRs or other rights that limit free access to plant genetic resources for food and agriculture or to their genetic parts or components in the form received from the multilateral system.

Thus, the treaty is consistent with international intellectual property (IP) standards, but it sets out a particular model for the acquisition of IP.

Those terms are also reflected, of course, in the SMTA and are passed on contractually to each recipient of genetic material from the gene pool. Then, if a recipient commercializes a product that is itself a plant genetic resource and that incorporates material from the multilateral system, and where such product is not available without restriction to others for further research, training, and breeding, the treaty specifies that the recipient shall pay 1.1% minus 30% of the sales of the commercialized product into the trust fund established by the governing body for the purpose of benefit-sharing (Bhatti, 2011). There is, consequently, a patent-based, benefit-sharing trigger here. The treaty also calls for the governing body to review the operation of this entire system and the terms, for access and benefit-sharing, 5 years after coming into force. It also states that the authority (governing body) may review the levels of payment in order to achieve fair and equitable sharing of benefits and may also assess whether the mandatory payment requirement in the material transfer agreement shall apply where commercialized products are available without restriction to others for further research and breeding program. This exercise was originally scheduled to be done in 2012, but it was postponed to the next session of the governing body. Research results were also governed by the SMTA, which requires that the recipient of material shall make available to the multilateral system through a global information system that result from research and development carried out on the material. The recipient was also encouraged to share, through the multilateral system, any non-monetary benefits that result from research and development. Finally, the SMTA provided that after the abandonment of IPRs, the product that incorporates the material should be placed back into one of the collections that are part of the multilateral system (Helfer, 2010; Bhatti, 2011).

3.9 SUSTAINABLE ACCESS TO COPYRIGHTED DIGITAL INFORMATION WORKS IN DEVELOPING COUNTRIES

Approaching "IP from below" elaborates the needs of users in both developed and developing countries for knowledge goods that are easy to get and having low price, especially for purposes of basic human development. "IP from below" promotes a bottom-up approach to innovation capacity building, especially for global sectors that are not technologically privileged. A top-down approach to capacity building in IP, by contrast, focuses on building capacity to comply with international IP's minimum standards, which in turn are thought to generate domestic innovative capacity through foreign direct investment, licensing, and technology transfer (Gervais, 2005). The TRIPS Agreement also refers development in its preamble, as well as in Articles 7 and 8. Moreover, the World Intellectual Property Organization (WIPO) founding agreement with the United Nations (UN) includes language regarding the need to "facilitate the transfer of technology related to industrial property to the developing countries in order to accelerate economic, social and cultural development." Similarly, Pamela Samuelson has inferred from the preamble of the WIPO Copyright Treaty (WCT) intent to preserve the traditional IP balance within global digital copyright that was already present within the Berne Convention framework, for purposes of education (Bartow, 1998; Arewa, 2006; Gerhart, 2007).

Whether through economic treaties such as the WTO or through the UN MDGs, distributionally fair social welfare gains will take place only when norms are both in the interests of the less powerful and more powerful actors (Bell, 1980). Thus, "IP from below" would highlight rather than footnote the perspectives of developing countries and, importantly, the non-elite users and consumers of knowledge goods within both developed and developing countries (Gerhart, 2007). This approach also overlaps with many prevailing critiques of IP maximalism. National governments may not represent the public interest; an approach from below views social movements and non-governmental organizations as relevant legal. "IP from below" also explores the practices of everyday resistance, such as "piracy" 21 or appropriation, rather than automatically demonizing them. And it places high value on democratic participation and decision-making, although in the context of global IP, most of the scholarly proposals thus far have focused on procedural rather than substantive reforms (Drahos and Braithwaite, 2002). However, a key difference between an approach from below and other critiques of the current IP balance is its emphasis on global distributive justice outcomes. The perspectives and actions of the least empowered among us ought to be included in more than just a formal equality sense in shaping a normative legal agenda. An approach from below clearly shapes IP outcomes with respect to knowledge goods by specific groups; in this case, users in developing countries for specific aims include novelty, access, and affordability. At least for purposes of this article, these goals also include basic human development as defined by the MDGs (Chon, 2006). The global IP framework poses distributive justice choices with far different inputs in order to create decision-making than on the domestic level. For developing countries, the effect of higher prices for global knowledge goods may be easier to distinguish than the relative impact for consumers in developed countries. Thus, policy choices will appear disproportionately to affect states with smaller markets, less international negotiating power, smaller budgets for public research, and poorer and less empowered consumers. But even in developed countries, which can more easily bear potential distributional burdens, the ongoing domestic debate on whether copyright law has over-privileged the author and submerged the user is one that goes squarely to the question of distribution. The globalization of IP sharpens distributive choices within all countries, especially in the context of digital networked technologies. Within the global framework of TRIPS, the articulation of a possible user right was one of the earliest signs recognizing the proper distribution between producer and user claims to value in public goods (Dinwoodie and Dreyfuss, 2006).

Digital technology has tremendous potential to control information for development. The recent appearance of the $100 hand-cranked laptop, run on open-source software, lends itself to countless possibilities for nontext-book-based distance education. The WIPO Copyright Treaty (WCT) does not exclude the performance of further domestic exceptions and limitations to digital rights sounding in copyright. There is currently an effort in WIPO, spearheaded by Chile, to study international minimum exceptions and limitations for educational and other uses in this context. There are strong efforts by the copyright content industries in developed countries to expand digital rights. Furthermore, WCT signatories are enacting technological protection measures required by Article 11, such as the arguably draconian US Digital Millennium Copyright Act (DMCA). These multilateral efforts have generated bilateral offspring. For developing countries, any additional ratcheting up of protections in the digital environment "arguably constitute a dead weight loss on already fragile economies: and should be viewed skeptically under a substantive equality paradigm (Dinwoodie, 2004). Instead, the essential public goods nature of information should be viewed as a potential development asset. An "IP from below" approach views the potential for diffusion and distribution of digital knowledge at almost zero marginal cost (once infrastructure is established) quite differently. These characteristics may be used to foster and develop the basic literacy and educational capacity that are prerequisites to the formation of a functioning future copyright content market. Especially where the danger to copyright interests associated with mass distribution via digital networks is reduced (e.g., because the work is culturally specific or is in a language that is not widely read), networked digital technology can be linked to diffusion models of information access. Countries should enact digital-specific educational exceptions where these are relevant and appropriate to their educational development policies. Possibly, these exceptions may even exceed the scope of Berne Article 10(2). Open course content initiatives in the tertiary textbook arena indicate that market-based mechanisms for distribution are only one possible means for providing access to textbooks. Intergovernmental organizations and prestigious educational institutions are now providing content without charge. Private–public partnerships for library digitization projects are proliferating. These and other new digital initiatives have enormous potential to expand the informational universes of educational institutions (Vaidyanathan, 2007).

3.10 CONCLUSIONS

There has been a dramatic increase in community-based interest in the last 10 to 15 years, from traditional commons managing the use of exhaustible natural resources by fixed numbers of people within natural borders, to global information commons, dealing with non-rival, non-excludible goods by a potentially limitless number of unknown users. The emerging global genetic resource commons fits somewhere in between, shifting in the direction of information commons as digital information infrastructures allow physically distributed commons to be networked in virtual global pools.

The analysis of a selected set of cases in this article shows that networking pools of genetic resources in a global commons potentially is a workable alternative to market-based solutions, which have been shown to be unable to generate sufficient investment in the vast quantities of genetic resources that are neglected because of their unknown and/or unlikely commercial value. These neglected resources are the building blocks for future scientific research and have enormous value for sustaining biodiversity and local livelihoods in developing and industrialized countries. Research and breeding require access to these multiple inputs that are combined into new compounds or screened to find organisms with new properties.

In the current legal environment, the range of obstacles to the full realization of the new opportunities offered by global networking of genetic resources presents a difficult challenge. This shows the need for suitable organizational forms, legal arrangements, and social practices, which can help to secure the global user community's need to address issues of common concern, such as global food security, global health, human development, biodiversity conservation, and climate change. As discussed in this chapter, in response to this challenge, governments, nonprofit organizations, global research communities, and breeders have developed a range of initiatives for the exchange of materials and information. The key issue is how to build upon these initiatives, and to put the budding global genetic-resource commons on a solid institutional basis that will enable commons-based production to co-exist, whenever effective, with market-based and state-based contributions to collective goods.

KEYWORDS

- **DNA libraries**
- **genepool**
- **genetic conservation**
- **germplasm**
- **plant genetic resources**

REFERENCES

Arewa, O. B., (2006). 'Piracy, biopiracy and borrowing: Culture, cultural heritage and the globalization of intellectual property', case research paper series in legal studies, Working Paper No. 4–19.

Bartow, A., (1998). '*Educational Fair use in Copyright*: Reclaiming the right to photocopy freely', University of Pittsburgh Law Review, *60*(1), 149–230.

Beattie, A. J., Barthlott, W., Elisabetsky, E., Farrel, R., Teck Kheng, C., & Prance, I., (2005). New Products and Industries from biodiversity, In: Hassan, R., Scholes, R., Ash, N., (eds.), *Ecosystems and Human Well-being: Current State and Trends*, Island Press, Washington, *1*, 271–295.

Bell, D., (1980). 'Brown v. Board and the Interest Convergence Dilemma', *Harvard Law Review*, *93*(3), 518–533.

Benkler, Y., (2006). *The Wealth of Networks: How Social Production Transforms Markets and Freedom*. Yale University Press, New Haven, 1–351.

Bhatti, S., (2011). *The International Treaty on Plant Genetic Resources*. National Academies Press (US).

Blakeney, M., (2001). Intellectual Property Aspects of Traditional Agricultural Knowledge, IP in Biodiversity and Agriculture: Regulating, Sweet and Maxwell, London, 63–89.

Boyle, J., (2008). *The Public Domain: Enclosing the Commons of the Mind*. Yale, New Haven, 1–340.

Braudel, F., (1992). (First published 1979). *Civilization and Capitalism*, 15th–18th Century, *The Structure of Everyday Life*. University of California Press, Berkeley, 1–250.

Burhenne-Guilmin, F., (2008). Biodiversity and international law: Historical perspectives and present challenges: Where do we come from, Where are we going? In: Jeffery, M. I., Firestone, J., Bubna-Litic, K., (eds.), *Biodiversity Conservation, Law + Livelihoods*, Bridging the North-South Divide. Cambridge University Press, Cambridge (UK), 26–42.

Byerlee, D., (2010). Crop improvement in the CGIAR as a global success story of open access and international collaboration. *International Journal of the Commons*, *4*(1), 452–480.

CBD, (2002). Bonn guidelines on access to genetic resources and fair and equitable sharing of the benefits arising out of their utilization. Secretariat of the Convention on Biological Diversity, Montreal.

CGIAR, (2003). Policy instruments, guidelines and statements on genetic resources, biotechnology and intellectual property rights, version II, produced by the system wide genetic resources programme (SGRP) with the CGIAR Genetic Resources Policy Committee, Rome.

CGIAR, (2009). CGIAR Joint Declaration, 8th December, Washington DC.

CGRFA, (2009). The use and exchange of animal genetic resources for food and agriculture. background study paper of the CGRFA, *43*, 55.

Chen, Y. F., & Liao, C. C., (2004). Intellectual property rights (IPR) for a biological resource center as the interface between academia and industry, In: *KurtböMicrobial Genetic Resources and Biodiscovery*, WFCC and University of the Sunshine Coast, Queensland, 1–350.

Chon, M., (2006). 'Intellectual property and the development divide', *Cardozo Law Review, 27*(6), 2821–2912.

David, P. A., (2008). The historical origins of 'open science': An essay on patronage, reputation and common agency contracting in the scientific revolution. *Capitalism and Society, 3*(2), Article 5, 1–106.

Dawyndt, P., Dedeurwaerdere, T., & Swings, J., (2006). Exploring and exploiting microbiological commons: contributions of bioinformatics and intellectual property rights in sharing biological information. *International Social Science Journal, 188*, 249–258.

Deci, E., (1976). The hidden costs of rewards. *Organizational Dynamics, 4*(3), 61–72.

Dedeurwaerdere, T., (2005). From bioprospecting to reflexive governance. *Ecological Economics, 53*(4), 473–491.

Dedeurwaerdere, T., (2011). *Beyond Patents: Collective Intellectual Property Strategies for the Conservation and Sustainable Use of Communities' Livestock, Crop and Microbial/Genetic Common Heritage.* Paper presented at the IASC 2011 Conference, Hyderabad, 1–26.

Dedeurwaerdere, T., Iglesias, M., Weiland, S., & Halewood, M., (2009). Use and exchange of microbial genetic resources relevant for food and agriculture. CGRFA background study paper no. 46. *Commission on Genetic Resources for Food and Agriculture*, Rome, 1–23.

Dedeurwaerdere, T., Krishna, V., & Pascual, U., (2007). An evolutionary institutional approach to the economics of bioprospecting, In: Kontoleon, A., & Pascual, U., (eds.), *Biodiversity Economics*: *Principles, Methods, and Applications*. Cambridge University Press, Cambridge (UK), 417–445.

Deek, F. P., & McHugh, J. A., (2008). *Open Source: Technology and Policy*. Cambridge University Press, Cambridge (UK).

Dinwoodie, G. B., & Dreyfuss, R. C., (2006). 'Patenting science: Protecting the domain of accessible knowledge', In: Guibalt, L., & Hugenholtz, P. B., (eds.), Th*e Future of the Public Domain*: *Identifying the Commons in Information Law*, 191–222.

Dinwoodie, G. B., (2004). 'Private ordering and the creation of international copyright norms: the role of public structuring', *Journal of Institutional and Theoretical Economics, 160*(1), 161–180.

Doyle, M., Jaykus, L. A., & Metz, M., (2005). *Research Opportunities in Food and Agriculture Microbiology*, Report from the American Academy of Microbiology, American Academy of Microbiology, Washington, DC, 710–727.

Drahos, P., & Braithwaite, J., (2002). Information feudalism: who owns the knowledge economy, *Earthscan*, London.

Dreyfuss, R. C. (2004). 'TRIPS-Round II: Should Users Strike Back?,' *University of Chicago Law Review, 71*(1), 21–35.

FAO, (2007): The state of the world's animal genetic resources for food and agriculture – in Brief, edited by Dafydd Pilling & Barbara Rischkowsky. FAO, Rome.

Fowler, C., Smale, M., & Gaiji, S., (2001). Unequal exchange? Recent transfers of agricultural resources and their implications for developing countries. *Development Policy Review, 19*, 181–204.

Frey, B., & Jegen, R., (2001). Motivation crowding theory. *Journal of Economic Surveys*, 15(5), 589–611.

Gerhart, P. M., (2007). 'The tragedy of TRIPS', *Michigan State Law Review, 1*, 143–184.

Goeschl, T., & Swanson, T., (2002). The diffusion of benefits from biotechnological developments: The impact of use restrictions on the distribution of benefits, In: Swanson, T., (eds.), *The Economics of Managing Biotechnologies.* Kluwer Academic Publishers, Dordrecht, 219–250.

Goff, S. A., (2011). The plant collaborative: Cyber infrastructure for Plant Biology. *Front. Plant Sci.*, *2*, 34.

Halewood, M., & Nnadozie, K., (2008). Giving priority to the commons: The international treaty on plant genetic resources, In: Tansey, G., & Rajotte, T., (eds.), *The Future Control of Food.* Earthscan, London, pp. 115–140.

Halewood, M., (2010). Governing the management and use of pooled microbial genetic resources: Lessons from the global crop commons. *International Journal of the Commons*, *4*(1), 404–436.

Hall, S. J. G., (2004). *Livestock Biodiversity*. Oxford, Blackwell Publishing Company, 1–320.

Hammond, W. N. O., & Neuenschwander, P., (1990). Sustained biological control of the cassava mealybug *Phenacoccus manihoti* (Hom.: Pseudococcidae) by *Epidinocarsis lopezi* (Hym.: Encyrtidae) in Nigeria. *BioControl*, *35*(4), 515–526.

Helfer, L. R., (2010). Using intellectual property rights to preserve the global genetic commons: The International Treaty on Plant Genetic Resources for Food and Agriculture, 217–224.

Helfer, L., (2005). Agricultural research and intellectual property rights, In: Maskus, K. E., & Reichman, J. H., (eds.), *International Public Goods and Transfer of Technology*. Cambridge University Press, Cambridge (UK), 188–216.

Hess, C., & Ostrom, E., (2007). Understanding knowledge as a commons: From theory to practice. MIT Press, Cambridge (MA).

Jeffery, M. I., (2008). Biodiversity conservation in the context of sustainable human development: A call to action, In: Jeffery, M. I., Firestone, J., & Bubna-Litic, K., (eds.), *Biodiversity Conservation, Law + Livelihoods*. Bridging the North-South Divide. Cambridge University Press, Cambridge (UK), 69–93.

Jinnah, S., & Jungcurt, S., (2009). Global biological resources: Could access requirements stifle your research? *Science*, *323*(5913), 464–465.

Kurtböke, I., & Swings, J., (2004). Microbial genetic resources and biodiscovery. Surrey (UK), *CABI Bioscience*, 1–400.

Lessig, L., (2001). The fate of the commons in a connected world, *The Future of Ideas*. 1–368.

Lessig, L., (2008). *Remix: Making Art and Commerce Thrive in the Hybrid Economy*. The Penguin Press, New York.

Mäki-Tanila, A., Tvedt, M. W., Ekström, H., & Fimland, E., (2008). *Management and Exchange of Animal Genetic Resources*, Norden, Copenhagen, 1–81

Matasci Data Access for the 1,000 Plants (1KP) project, (2014). *Giga Science*, *3*, 17.

Nguyen, T., (2007). Science commons: Material transfer agreement project. Innovations: *Technology, Governance, Globalization, 2*(3), 137–143.

Parry, B., (2004). *Trading the Genome*. Columbia University Press, New York.

Rai, A. K., (1999). Regulating scientific research: Intellectual property rights and the norms of science. *Northwest University Law Review, 94*, 77–152.

Reichman, J. H., Dedeurwaerdere, T., & Uhlir, P. F., (2009). Designing the microbial research commons: Strategies for accessing, managing and using essential public knowledge assets. Book manuscript presented at the national academies symposium on "*Designing the Microbial Commons*," Washington, 8–9.

Roa-Rodríguez, C., & Van Dooren, T., (2008). Shifting Common spaces of plant genetic resources in the international regulation of property. *The Journal of World Intellectual Property, 11*(3), 176–202.

Safrin, S., (2004). Hyper ownership in a time of biotechnological promise: The international conflict to control the building blocks of life. *American Journal of International Law, 98*, 641–685.

Simpson, R. D., Sedjo, R. A., & Reid, J. W., (1996). Valuing biodiversity for use in pharmaceutical research. *Journal of Political Economy, 104*(1), 163–185.

Smith, D., (2007). The implementation of OECD best practice in WFCC member culture collections, In: Stackebrandt, E., Wozniczka, M., Weihs, V., & Sikorski, J., (eds.), Connections between Collections. *WFCC and DSMZ, Goslar*, 18–21.

Stern, S., (2004). *Biological Resource Centers*. Washington, DC: Brookings Institution Press.

Sulston, J., & Ferry, G., (2003). *The Common Thread:* Science, politics, ethics and the human genome. Corgi Books, London.

Swanson, T., & Goeschl, T., (1998). The management of genetic resources for agriculture: Ecology and information, externalities and policies. *CSERGE Working Paper GEC, 98*, 12–27.

Swanson, T., & Goeschl, T., (2005). Diffusion and distribution: The impacts on poor countries of technological enforcement within the biotechnology sector, In: Maskus, K. E., & Reichman, J. H., (eds.), *International Public Goods and Transfer of Technology*. Cambridge University Press, Cambridge (UK), 669–694.

Tvedt, M. W., Hiemstra, S. J., Drucker, A. G., Louwaars, N., & Oldenbroek, K., (2007). Legal aspects of exchange, use and conservation of farm animal genetic resources. *Fridtjof Nansens Institute Report,* 1–41.

CHAPTER 4

TRADITIONAL KNOWLEDGE AND ITS PROMOTION THROUGH PROVIDING LEGAL RIGHTS

KUMARI RAJANI,[1] TUSHAR RANJAN,[2] RAVI RANJAN KUMAR,[2] GANESH PATIL,[3] MAHESH KUMAR,[2] JITESH KUMAR,[2] and MUKESH KUMAR[1]

[1] *Department of Seed Science and Technology, Bihar Agricultural University, Sabour–813210, India, E-mail: rajani.dsstiari@gmail.com*

[2] *Department of Molecular Biology and Genetic Engineering, Bihar Agricultural University, Sabour–813210, India*

[3] *Vidya Pratisthan's College of Agriculture Biotechnology, Vidyanagari, Baramati–413133, India*

CONTENTS

ABSTRACT

Traditional knowledge, embedded in the cultural traditions of regional/indigenous/local community, includes knowledge about traditional technologies of livelihood, midwifery, ethno, and ecological knowledge that are crucial for the subsistence and survival and are generally based on accumulations of empirical observation and interaction with the environment. In recent years, traditional knowledge has grown tremendously in view of its value to biotechnology, particularly the pharmaceutical, phytomedicinal, nutraceutical, and herbal sectors. Three-fourth of the biologically active plant-derived compounds currently in use have been discovered through follow-up research to verify authenticity of data derived from traditional sources. The conventional intellectual property law does not cover inventions and innovations of indigenous and local people. Their contributions to plant breeding, genetic enhancement, biodiversity conservation, and global drug development are not recognized, compensated, and even protected. Similarly, the traditional knowledge of indigenous and local people is not treated as intellectual property worth protection, while the knowledge of modern scientists and companies is granted protection. Thus, there should be an enormous and solid effort to protect traditional knowledge, and several legal frameworks are being adopted for their protection. Growing interest and catapulting markets in "natural" food, medicinal, agricultural, and body products signals increased research activities into traditional knowledge systems. The patentability of products and/or processes derived from traditional knowledge of indigenous and local people poses a number of critical questions associated with compensation for the knowledge and protection against future uncompensated exchange of the knowledge. The imbalances in the intellectual property system have been created and are sustained by established mechanisms of accessing the modern economic space and power. Indigenous and local people often experience insecure resource tenure, are financially weak, and lack institutional arrangements to safeguard their property rights. In this chapter, we have discussed a burning issue of protecting indigenous and traditional knowledge and practices. We have discussed that the traditional rights should be recognized and they should be given a decisive voice in formulating policies about resource development in their areas.

It further recommends that local institutions through which indigenous and local people socialize and conduct their economic activities should be strengthened. Conventional intellectual property law does not adequately cover or protect traditional knowledge and innovations of indigenous and local people. However, non-patent forms of intellectual property protection could be exploited to protect the knowledge and innovations.

4.1 INTRODUCTION

Traditional knowledge (TK)/indigenous knowledge/local knowledge basically involves investigating hidden knowledge embedded in the cultural traditions of regional, indigenous, or local community. Basically, it includes in detail about traditional technologies of livelihood (such as tools for bee keeping, techniques for hunting or agriculture), midwifery(sexual and reproductive health of woman), ethno botany (relationship between plants and local people), ecological knowledge (cultural relationship with the environment), celestial navigation (finding direction by observing the sun and moon), ethno astronomy etc. These kinds of awareness are critical for survival and are generally based on accumulations of empirical observation and interaction with the environment. Traditional knowledge is seen as being formed historically and expressed as ideas and practices, which range from spiritual to legal and scientific developed out of the traditional ways of living indigenous communities. However, in rapidly changing socioeconomic environment, TK represents a potential source of wealth to be employed for the help of both indigenous communities and the world as well (Quinn, 2001). The idea of TK and indigenous peoples has gotten overall consideration in international forum on sustainable development and additionally in intellectual property (IP) protection.

Johnson, an anthropologist, characterizes customary information as an assemblage of learning made by a gathering of individuals living in close cooperation with nature. The qualities of conventional learning include: (i) Creation of data and their trade from era to era; (ii) Continuous upgrade in light of the fact that new information is incorporated to the current one; and (iii) Both creation and change of information is a collective endeavor (Quinn, 2001). The above proclamation can be streamlined by the story below. For example, an Achuar (Jivaro) man chomped by a snake in a secluded zone of the Peruvian rain forest was given a snakebite cure by a bi-social Candoshi-Achuar man who knew about this cure from his mom's tribe. On drinking the preparation, the man felt relief from pain around the puncture site, perhaps due to reduced inflammation. On return to his community, he expounded the virtues of this "new" anti-venomous plant and on a return visit in 6 months, we discovered that this treatment had become generally adopted as part of the Achuar traditional pharmacopeia, all as a consequence of one man's experience. (Lewis et al., 1991; Quinn, 2001). In recent years, TK has flourished tremendously in view of its value to biotechnology, particularly the pharmaceutical, phytomedicinal, nutraceutical, and herbal sectors. Three-fourth of the biologically active plant-derived compounds currently in use have been discovered through follow-up research to verify authenticity of data derived from traditional sources. Scientists are involved to validate the importance of an ethno-botanically targeted approach to the initial discovery of therapeutics. Such research draws on the conventional learning of nearby and indigenous groups who have authority of such assets, thus permitting a focus on testing of particular plants for particular purposes (Farnsworth et al., 1985; Schuster, 2001). Figure 4.1 represents different branches, their transmission, and beneficial effect of TK.

Thus, there should be an enormous and solid effort to protect TK, and several legal frameworks are being adopted for their protection, such as, for example, (a) International security through arrangements and traditions, (b) National assurance through national enactments controlling access to hereditary material established in different nations, and national licensed innovation enactments, and (c) Local assurance through private legally binding measures, documentation, and their safeguarding (Dutfield, 2001). The Government of India has prepared a Traditional Knowledge Digital Library (TKDL) on TK about medicinal plants, toward off incidence of piracy, containing 36,000 formulations used in *Ayurveda*, *Unani*, *Siddha*—Indian system of medicine, from 14 ancient books. In India, TKDL, Central Scientific and Industrial Research, and department of AYUSH (Department of Ayurveda, Yoga and Naturopathy, Unani, Siddha and Homoeopathy) had begun numerous collaborative projects to look over this and the role of medicinal plants such as Turmeric and Neem in wound healing and antifungal, respectively, stimulated the formation

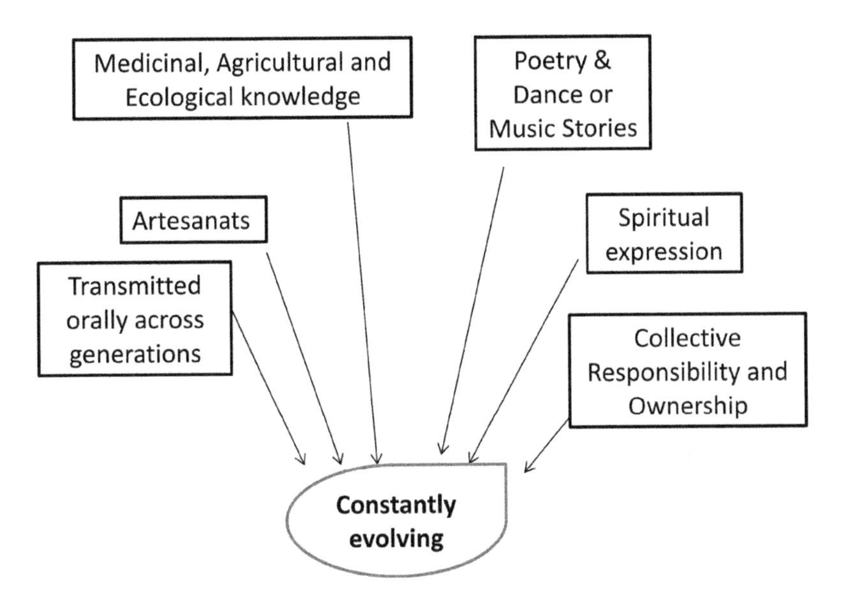

FIGURE 4.1 Represents nature, composition, evolution, benefits, and preservation of traditional knowledge (TK).

of TKDL. Recently, a study also proved the role of Jamun (Blackberry), eggplant, and bitter gourd in controlling diabetes. To the extent, indigenous information held and utilized by people who recognized themselves as indigenous of a place in light of a "blending of social peculiarity and earlier regional inhabitance in respect to an all the more as of late arrived populace with its own particular unmistakable and in this manner prevailing society." TK is maintained by members of a diverse culture and/or sometimes acquired "by means of investigating unique to that culture, and concerning the culture itself or the local environment in which it exists." Indigenous knowledge can be considered in the TK category, but TK is not always basically indigenous (Figure 4.2). TK is particularly the totality of all information and practices, regardless of whether unambiguous or certain, utilized as a part of the administration of financial and natural aspects of life. Traditional learning is basically the set up on past experiences and perception. It is typically a collective property of a general public. Several individuals from the diverse culture add to it after some time, and it is adjusted and expanded as it is utilized over the long haul (Brush and Sta-

FIGURE 4.2 The traditional knowledge system.

binsky, 1996). This knowledge is transferred subsequently from one generation to other. According to the United Nations Environment Program (UNEP), this learning can be diverged from sophisticated information to monetary inclinations and theories with those of other broad societies. It is for the most part the characteristics of a specific people, who are personally connected to a specific socio-natural situation through different financial, social, and religious activities.

TK is vibrant in nature and amends its nature as the necessities of the general population change. It likewise makes a choice of imperativeness from being intensely excavate in individuals' lives. It is hard to disconnect or file customary information from indigenous peoples. Cases of TK incorporate information about the utilization of particular plants or potentially parts thereof, identification of medicinal properties in plants, and gathering hones. There is adequate and growing evidence of TK and indigenous knowledge and related works on contributing altogether to the preservation and upgrade of biodiversity. Local people exemplifying traditional lifestyles and knowledge have devised and deploy various technologies to conserve the environment in general and biodiversity in particular (Mugabe, 1994; Juma and Ojwang, 1996; Moran et al., 2001).

Since the 1980s, the address of issues identifying with indigenous individuals by different nations of world has progressively focused on the possibility of TK. This recognition of TK on the international level is comparatively recent and began at the second half of the 19th century.

4.2 TRADITIONAL KNOWLEDGE AND BIODIVERSITY PROSPECTING

Modern biotechnology, pharmaceutical, and human social insurance ventures have expanded their enthusiasm for natural products as sources of new biochemical compounds for drug, chemical, and agro-items advancement since the last few decades. The resurgence of enthusiasm for conventional learning and drug can be seen in these decades. This conspiracy has been stimulated by the significance of TK as a lead in new article improvement. Out of 119 medications created from higher plants and on the world market today, it is assessed that 74% were developed from traditional herbal medicine. The annual world market for medicines derived from medicinal plants discovered from indigenous people amounted to US$ 43 billion in 1985.

The indigenous peoples residing in developing countries have contributed considerably to the global drug industry. Okoth-Owiro and Juma estimated that plant-derived prescription drugs in the USA originate from 40 species of which 50% are from the tropics. About US$4 billion economy in the USA itself is generated by the products of 20 plant species. TK has led to the search for these plants. The National Cancer Institute (NCI) of the USA invested in extensive collections of *Maytenus buchananii* from Simba Hills of Kenya in the 1970s. The NCI was generally led by the knowledge of the Digo communities indigenous to the Simba Hills area, who have used the plant to treat cancer since long time. More than 27.2 tons of the shrubs were collected by the NCI from a game reserve in the Simba Hills for testing under a major screening program. The plant yields maytansine was considered a potential candidate for the treatment of pancreatic cancer. Unfortunately, all the material collected was traded without the consent of the Digo community, neither was there any recognition of their knowledge of the plant and its medicinal properties (Mugabe, 1994). The plant species, namely *Homalanthus nutans,* contains the anti-HIV compound prostratin, and was also collected from the Samoa rainforests by NCI. The collection of this plant species was attempted on the ground of TK (Rafi, 1994). The NCI has also benefited from TK of local communities living around Korup Forest Reserve in Cameroon. The institute collected *Ancistrocladuskor rupensis* to screen for an anti-HIV principle, Michellamine B. This bio-prospecting effort has advanced into pre-clinical development. Based on the success and utilization of the TK, NCI and other drug innovative companies have invested extensive funds to prospect for plants containing valuable chemicals, and a number of them are examining the adequacy of traditional medicines. Although trade in medicinal plants from developing countries has increased in the past few decades with more drugs developed, barely any advantages accumulate to the source nations and the conventional groups. Entire trade in herbal remedies and botanicals in 1995 yielded over US$ 56 billion, and the only payments to the local communities were for the manual labor involved. According to Posey, even less than 0.001% of the profits from these drugs established from natural products and TK accrue to indigenous people who provided technical leads for the research. Shaman develops new therapeutics by working with indigenous people of tropical forests. The Body Shop is bio-prospecting in the Kayapo area of Brazil by extensively drawing on TK of the Kayapo Indians. It has invested in ethno-botanical research for the development of new ingredients for its body-care products. In 1991, the Body Shop had at least 300 products with annual sales of US$90 million. By 1995, its annual sales stood at least at US$ 200 million (Juma, 1989; Posey, 1991).

Rice-Tec, a US organization, has been offering Basmati rice developed in the USA under the trademark Texmati and Kasmati since 1980s. Texmati conveys the depiction "American-style Basmati rice," while the unrivaled Kasmati is portrayed as "Indian-style Basmati rice." In December 1995, at the initiative of the Indian government, a Basmati Development Fund was set up to protect the Basmati trademark. Battles have been won by the Agricultural Products Export Development Authority (APEDA) and its lawyers, from the Delhi firm of Kumaran and Sagar in countries such as the UK, Greece, Colombia, Brazil, and Spain. Rice-Tec, which used a number of Indian scientific publications in its support, was granted a US patent on September 2, 1997. India filed a re-examination request for the patent on Basmati rice lines and grains (US Patent No. 5,663,484) granted by the USPTO, and Rice-Tec Company from Texas has decided to withdraw the specific claims challenged by India (Mashelkar, 2001).

In general, a huge piece of the worldwide economy depends on the appropriation and use of traditional knowledge. In reality, conventional information is progressively adding to creation in current economies where property rights are inimical to community IP. Modern economic policies and laws (particularly modern property laws) undervalue this knowledge: at best they ignore it and at worst they contribute to its destruction. Indigenous people and their knowledge are, however, vulnerable to the destruction. Recent estimates show that 85 Brazilian–Indian groups became extinct. In the Amazonian region, on an average, one Amerind group has become extinct each year. The destruction of indigenous people, indigenous communities, and their knowledge is caused by many interrelated and complex factors. This may include destruction of ecosystems in search for expanded agricultural lands, deforestation associated with harvesting of timber and other forest products, and appropriation of TK with no rewards to the holders of that knowledge. Concern over the growing economic interests and economic importance of TK as well as the loss of this knowledge has generated a wide range of public policy issues including those associated with IP protection. Growing interest and catapulting markets in "natural" food, medicinal, agricultural, and body products signals increased research activities into traditional knowledge systems. Now, more than ever, the intellectual property rights (IPR) of native people must be protected and compensation for knowledge guaranteed (Juma, 1989; Posey, 1991).

4.3 TRADITIONAL KNOWLEDGE AND INTELLECTUAL PROPERTY RIGHTS

IP laws have recently received prime attention for technological innovation and industrial field. It is also witnessed as a tool for promoting the conservation of biological diversity, sustainable use of its components, ensuring that benefits arising from the utilization of genetic resources are shared in a fair and equitable manner among the stakeholders. According to critics, IP protection increases the costs of products, promotes genetic monoculture by concentrating industrial, and agricultural activities on a few cultivated varieties or species, and, when extended to plants and animals, is in conflict with the morals of many societies. IP laws vary in nature and scope from country to country. The IP protected in one country may not be recognized in another country. Even though there are several international agreements to harmonize IP protection, many differences among national laws, especially regarding patents still persists. For example, USA and European Union countries allow patent protection over genetically engineered organisms, which meet the normal requirements for patentability, while many other countries are opposed to extending patents to such subject matter. There are also differences in the duration of patent protection from one country to another for which an inventor is granted a patent. In addition, different countries have different conditions for the disclosure of information concerning the invention. While some (like USA and the European Union countries) have strict rules and mechanisms for enforcing patent application requirements, others (particularly developing countries) have weak institutional arrangements for ensuring compliance with disclosure requirements (Greaves, 1994).

These differences in national application of IP law are at the center of much of the debate on the IPR of indigenous and local people. The case of TK of indigenous and local people has opened debate on the competence and ethics of IP protection. The debate, particularly the absence of consensus on whether and how to extend IP protection to TK on these issues is complex and controversial. This is partly because of differences in conceptual treatment and often lack of clarity on the two concepts of TK and IP. It is further because of scant advantage of impression is accessible to the governing body responsible for policy and behavior making at both national and international levels. In addition, these issues are often debated in United Nations, nongovernmental organizations, conferences, and other social platforms, each with its distinct sectoral interest and focus in the subject. For example, dialog (between ILO and the United Nations Working Group on Indigenous Populations) on the human rights of indigenous people has rarely addressed. The World Trade Organization (WTO) has not confronted the implications of its Agreement on Trade-

Related Aspects of Intellectual Property Rights (TRIPS Agreement) for the protection and use of TK. In general, international debate, on issues of IP protection and rights in TK in particular, is characterized by apprehension and discrepancy. However, environmental NGOs, anthropologists, and the Convention on Biological Diversity (CBD) have begun to create a strong political foundation for addressing these issues in a holistic manner. The CBD's holistic nature and its large and diverse constituency open to NGOs have provided an inter-governmental forum where these issues are being debated with a certain measure of coherency (Moran et al., 2001).

The debate in the CBD and other forums now swing between two extremes: one advocates the extension of IP protection to cover TK, even including patenting, and another that promotes the status quo where such knowledge is treated as a public good. Those who advocate the extension of IP protection to cover TK often advance the following arguments. First, the extending of IP protection to TK would promote technological innovation for facilitating the dissemination and development of that knowledge in the modern economic space. Second, incentives can be generated for local and indigenous people to conserve the environment and manage biodiversity, where IPR in TK get recognition. Third, the industrialized countries have a moral responsibility to ensure that indigenous and local people receive an equitable share of benefits arising from the use of their TK and commercialization of genetic resources (Agrawal, 1995).

It is crucial to remember that the underlying aim of IPR is to crook noesis into a marketable commodity, not to conserve such knowledge in its most-fitting cultural context of use. This end necessarily translates into a focus on segregating and isolating information into identifiable and manageable pieces that can be protected by law as IP. In contrast, ethnobotanical knowledge by its very nature is integrative, holistic, and synergistic. It is most meaningful in situ where industrial plants are understood in relation to the ecological and cultural environments in which they have been grown, managed, and used by topical anesthetic residents. IPR depart from such tradition by valuing the discrete properties of plants that can most easily be taken out of their natural and cultural contexts and replicated through artificial selection in a laboratory or greenhouse. Given the legitimate premises on which IPR are based, it is improbable that IPR will ever be a valuable model for protecting ethno-botanical knowledge.

4.3.1 THE PARIS CONVENTION FOR THE PROTECTION OF INDUSTRIAL PROPERTY

The Paris Convention for the Protection of Industrial Property, held on March 20, 1883, is an international legally binding agreement concerning property rights in patents, utility models, industrial designs, service marks, indications of source, or appellations of origin and trademarks. The Convention has, as on March 2017, 177 contracting member countries. Article 1 of the Convention highlights the scope of industrial property. It is stated in Article 1(3) that "industrial property shall be understood in the broadest sense and shall apply not only to industry and commerce proper, but also to agricultural and other manufactured or natural products, like; grain, tobacco leaf, fruit, wines, beer, cattle, mineral waters, minerals, flowers, and flour." Article 2 set conditions for national treatment that each Contracting Party to the Convention must concede a similar IP protection to nationals of other Parties that it gives to its own particular nations.

Article 5(a) of the Convention allows Parties to pass legislation to grant compulsory licenses in order to prevent abuses resulting from the activities of selected rights. It is possible for improvement of indigenous and local people to be protected under the trademark, utility models, industrial designs, service marks, and indications of source or labeling of origin provisions of the Paris Convention. In this regard, Article 7 of the Convention is significant and worth noting. It enables member countries to "accept for filing and to protect collective marks belonging to associations the existence of which is not opposing to the law of the country of origin, even an industrial or commercial establishment is not possessed by such association." If indigenous and local people form associations that are legally legitimate in their countries, it is possible for them, as a collectivity, to acquire collective marks. This Convention recognizes and protects modern industrial products and services generated from that information, but the convention does not contain provisions for granting patents to traditional knowledge per se or any other kind of knowledge for that matter (Goldstein et al., 1997).

4.3.2 PLANT BREEDERS' RIGHTS

Plant breeders' rights cover the sovereign rights on plant varieties developed by the breeders. They imply exclusive exploitation rights in the developers of new plant varieties as an incentive to pursue innovative activity and to enable breeders to recover their investment. Like most intellectual property rights, plant breeders' rights are provided for certain time period, at the end of which the varieties pass into the public domain. UPOV (International Convention for the Protection of New Varieties of Plants), 1978 and 1991, established minimum international standards for the protection of plant breeders' rights.

Plant breeders' rights, under the UPOV Convention, provide intellectual property protection to plant varieties that are novel, distinct, uniform, and stable. The terms and conditions are similar to those for patenting, although the requirements of "novelty" and "distinctiveness" for purposes of plant breeders' rights are interpreted more compassionately than for patent protection. Plant breeders' rights are useful regimes for those countries that do not wish to extend patents to plant varieties and other living organisms. However, in UPOV 1991, several modifications were made that tilt plant breeders' rights more toward patents. The first was the expansion of subject matter for protection under the regime of plant breeders' rights. The 1978 Act of the UPOV Convention provided protection only to plant varieties of nationally defined species. However, the 1991 Act extends protection to varieties of all genera and species. In addition, the revised UPOV Convention has extended protection to commercial use of all materials of the protected variety, while the 1978 regime restricted the commercial use of only the reproductive material of the variety. Second, the "farmer's privilege" in the 1978 Act was more limited than the 1991 amendment, under which it was left to Member States of UPOV to determine on an optional basis whether to exempt from the breeder's rights any traditional form of saving seed. Under UPOV 1991, a farmer, who produces a protected variety from farm-saved seeds, is guilty of breaching the code unless the national law provides otherwise. This weakens the economic position of rural farmers and suppresses local and traditional innovations. Additionally, no provisions for recognizing the contribution and knowledge that indigenous and local people implement during plant breeding programs were mentioned in UPOV convention. In our view, therefore, plant breeders' rights as embodied in the 1991 Act of the UPOV Convention are inadequate in protecting traditional knowledge of indigenous and local people (Agrawal, 1995).

4.3.3 PROTECTION OF TRADITIONAL KNOWLEDGE UNDER TRIPS

The negotiation and adoption of the TRIPS Agreement as part of the Uruguay Round in 1994 have added new dimensions to the debate on intellectual property rights in traditional knowledge. The TRIPS Agreement sets minimum standards for countries to follow in protecting intellectual property. Its main objective is "to reduce deformations and obstacles to international trade, and taking into account the need to promote effective and adequate protection of intellectual property rights, and to ensure that measures and procedures to enforce intellectual property rights do not themselves become barriers to legitimate trade." The countries who approved this Agreement established comprehensive intellectual property protection systems covering patents, copyright, geographical indications, industrial designs, trademarks, and trade secrets. However, Article 1 of the TRIPS Agreement (on the nature and scope of the obligations) provides some flexibility in the implementation of the provisions of the Agreement. It states in paragraph 1 of the Article that "members may, but shall not be obliged to, implement in their domestic law more extensive protection than is required by Agreement, provided that such protection does not contravene the provisions of Agreement." According to Dutfield (1997), parties to the TRIPS Agreement can invoke this provision to enact legislation for protecting traditional knowledge. He asserts "the absence of any mention of traditional knowledge in the Agreement does not prevent any Member from enacting legislation to protect such a category of knowledge." After reviewing the TRIPS Agreement, we consider that it is not possible to protect TK under the current patent law. The TRIPS Agreement requires Member States to provide patent protection for "any inventions, whether products or processes, in all fields of technology, provided that they are new, involve an inventive step and are capable of industrial application." The "inventive step" and "capable of industrial application" requirements are deemed "to be synonymous with the terms 'non-obvious' and 'useful' respectively." TK products fail the test for patenting on one, or all, of the "new," "inventive step," and "industrial application" standards. On the new standard, they will probably fail because by its very nature traditional knowledge has been known for some length of time. One could try and argue that TK is new to the world outside of the community from which it came, but this is unlikely to succeed (Dutfield, 1997).

Article 29.1 of the TRIPS Agreement requires that a patent applicant disclose sufficient and clear information regarding the invention so that another person "skilled in the art" would be able to reproduce the product or complete the process. This is a standard patent law condition. Opponents of patenting have been quick to point out that this condition of information disclosure could erode the rights of indigenous and local people because it would make traditional knowledge easily available to commercial entities. Given the absence of financial and organizational competencies of indigenous and local people to monitor and enforce patents in modern economic space, their knowledge could easily be used without due compensation. The TRIPS Agreement does not enable the patenting of TK and/or traditional innovations due to its present situation. Article 27.2 of TRIPS states that "members may exclude from patentability inventions, the prevention within their territory of the commercial exploitation of which is necessary to protect order public or morality, including to protect human, animal or plant life or health or to avoid serious prejudice to the environment, provided that such exclusion is not made merely because the exploitation is prohibited by domestic law." The notions of *ordre public* (public order) and morality are not defined in the Agreement. However, it is clear that those inventions that cause injury to human, animal, and plant life as well as the environment may be excluded. States are given flexibility to adjudicate. Some may still provide patent protection for inventions that cause damage to the environment. Patenting of genetically engineered organisms and life forms is generally possible under these provisions. Further, patent protection may be provided by the country for a modified gene or a whole organism, which meets the normal requirements for patentability. Article 27.3(b) of the TRIPS Agreement has generated controversy and opportunity. It states that "other than non-biological and microbiological processes, members may also exclude from patentability, plants and animals other than microorganisms, and essentially biological processes for the production of plants or animals. However, members shall provide for the protection of plant varieties either by patents or by an effective *sui generis* system or by a combination thereof. The provisions of this sub-paragraph shall be reviewed four years after the entry into force of the WTO Agreement." First, there is controversy on what "an effective *sui generis*" regime is. "Effectiveness" of the *sui generis* system is not defined. The nature of a *sui generis* system is also left to individual members to determine. According to Crucible Group report of 1994, the term *sui generis* offers a wider range of policy choices as it presumably include any arrangement for plant varieties that offers recognition to innovator with or without monetary benefit or monopoly control. If there is any dispute on the nature and minimum standards of "an effective *sui generis*" system, the WTO is itself the mechanism for adjudication. Second, it has also been noted that multinational companies and developed countries are likely to promote plant breeders' rights as the effective *sui generis* system. "Plant breeders' rights may be used as a measure of effectiveness under the TRIPS Agreement, thereby limiting the ability of developing countries to develop a system to properly reflect their own social and economic needs." They will require or encourage developing countries to establish the UPOV arrangement. This, as Johnston and Yamin have rightly observed, could potentially remove plant varieties from the scope of the CBD and may significantly undermine the rights of local farmers. It could also erode prospects of ensuring that benefits from the use of plant genetic resources are shared in a fair and equitable manner. The TRIPS Agreement has, on the other hand, generated new opportunities to develop alternative property rights regimes that are ethically, socially, and environmentally appropriate to the needs and conditions of indigenous and local people in developing countries. As stated earlier, under Article 27.3(b) of the TRIPS Agreement, Members may establish effective *sui generis* regimes. This is an opportunity which developing countries should quickly tap by devising and promoting nonpatent measures. They could easily lose out if Article 27.3(b) were to be removed from the Agreement during its review in 1999. Some developed countries, particularly the USA, are already campaigning for its removal so that no restrictions are imposed on the patenting of life forms. The TRIPS Agreement itself does not provide any protection for the traditional knowledge and innovations of indigenous and local people but it creates flexibility for establishing alternative non-conventional intellectual property protection measures. The non-conventional can be represented in some form and can distinguish the product or service which it is representing. For example, an unconventional trademark is a new type of trademark which does not fall into the category of conventional or traditional trademarks. These trademarks fulfill the conditions of being a trademark but are difficult to register because of their unusual nature.

A conventional intellectual property law does not cover traditional inventions and innovations of indigenous and local people. Their contributions to biodiversity conservation, plant breeding, genetic enhancement, and global drug development are not recognized, compensated, and even protected. The knowledge of modern scientists and companies is granted protection under IP laws but at the same time, the TK of indigenous and local people is not treated as IP worth protection. Critical questions arise on the patentability of products and/or processes derived from TK of indigenous and local people and compensation for the knowledge, and protection against future uncompensated exchange of the knowledge. Due to this, partiality in the IP system is being created and is sustained by established mechanisms of accessing the modern economic space and power. Indigenous and local people often experience insecure resource tenure, and thus, the issues extend to fundamental and more complex questions of human rights of these people (Dutfield, 1997; Goldstein, 1997).

4.4 TRADITIONAL KNOWLEDGE AND RIGHTS OF INDIGENOUS PEOPLE

The debate on the protection of TK through IP law has recently moved to the human rights forums for several reasons. First, the appropriation of the knowledge by industrialized country firms and researchers without fair compensation or reward to indigenous and local people is now seen as breaching fundamental ethics, moral, and legal norms that protect people from any form of economic, ecological, political, and social abuse. Second, knowledge of indigenous and local people is their property, and there is no reason why international law should discriminate against them and create barriers to their enjoyment of the rights in that property.

The concern in the human rights forums is therefore whether and how to apply international human rights standards and laws to protect TK of indigenous and local people as their IP. Existing international and national laws and programs do not clearly recognize rights in TK as part of the bundle of human rights. The Universal Declaration of Human Rights 1948 (the UDHR) and the International Covenant on Economic, Social and Cultural Rights 1966 (the ICESCR) contain provisions that could be interpreted to cover rights of indigenous and local people.

For example, Article 1 of the ICESCR establishes the right of self-determination, including the right to dispose the natural wealth and resources. It also states the right to protect and conserve resources, including intellectual property. It was argued that intellectual property to the traditional knowledge of indigenous people can be extended using Article 7 of the UDHR. It is stated in Article 7 that "All are equal before the law and are entitled without any discrimination to equal protection of the law. Equal protection will be provided to all against any discrimination in violation of this declaration and against any stimulation to such discrimination. The protection of traditional knowledge of indigenous and local people as well as demand for the sharing of benefits arising from the use of that knowledge could be appealed. Article 27.1 states on the right of participation for all in the cultural life of the community, to enjoy the arts and to share in scientific advancement and its benefits." This Article 27.1 also provides a legal basis on the benefits arising from the use of their knowledge and resources for indigenous and local people. Article 27.2 states about the right of protection of the moral and material interests resulting from any scientific, literary or artistic production of which a person is the author. Indigenous and local people have moral, cultural, and material interests in their traditional knowledge, and thus, these interests should be protected by protecting their products and knowledge. Overall, the UDHR contains provisions on a wide range of civil, political, economic, social, and intellectual rights. As already observed, it is Article 27 of the Declaration that is particularly relevant to the issue of intellectual property protection of traditional knowledge. However, there are several limitations to use it as a legal instrument for protecting traditional knowledge of indigenous and local people. First, while traditional knowledge is a collective property and generates collective rights, the UDHR majorly covers individual rights (Posey, 1994).

The International Labor Organization (ILO), the first United Nations agency, addressed the issues of indigenous people' rights. In 1926, ILO established an expert committee to develop international standards for the protection of native workers. This committee generated the basis for the adoption, in 1957, of the convention concerning the protection and integration of indigenous and other tribal and semi-tribal populations in different countries. This convention commonly referred to as convention 107, which mainly focused to incorporate the indigenous people into modern production systems. This convention was revised in June 1989 as convention 169 concerning indigenous and tribal people in independent countries. The revised Convention deliberately avoided using the approach of promoting the integration of indigenous and tribal people. It promoted the protection of indigenous people as distinct and separate people. Article 2.2(b) provides that governments shall have the responsibility of developing measures for "promoting the full realization of the social, economic and cultural rights of these people with respect for their social and cultural identity, their customs and traditions and their institutions." Article 5(a) provides that "the social, cultural, religious and spiritual values and practices of these people shall be perceived and protected, and due account shall be taken of the nature of the problems which confront them as groups and individuals." These provisions should be broadly read to include recognition and protection of traditional knowledge of the people. Convention 169 also contains provisions that explicitly recognize collective rights of indigenous people. For example, Article 13.1 states that "governments shall respect the importance of the cultures and spiritual values of the people concerned of their relationship with the lands or territories, or both, which they occupy or use, and in particular the collective aspects of this relationship." This provision provides a basis of argument for the enlargement of IP regimes to accommodate collective rights of indigenous people. However, the convention has not been adequately raised to create the legal basis for creating intellectual property rights in traditional knowledge of indigenous people (Axt et al., 1993). The adequacy of Convention 169 is a concern of some indigenous groups and NGOs with a number of the provisions of the Convention. First, the convention only requires that indigenous people be consulted on matters affecting them. It does not require that the consent of these people be sought before measures affecting them are instituted. Second, the groups are of the view that provisions dealing with land and natural resources are insufficient.

The rights of indigenous people have also been spoken in the United Nations Economic and Social Council. In 1972, the Council established under its Commission on Human Rights a Sub-Commission on the Prevention of Discrimination and Protection of Minorities. The Sub-Commission commissioned a study on discrimination against indigenous populations. The study, completed in 1983, concluded that current human rights standards are not fully applied to indigenous people and those international legal instruments are not "wholly adequate for the recognition and promotion of the specific rights of indigenous populations as such within the overall societies of the countries in which they now live." It was recommended to adapt the declaration of the convention. In addition, the sub-commission recommended the establishment of a working group on indigenous populations to: (1) "Review developments leading to the support and protection of the human rights and fundamental independence of indigenous populations and; (2) Give special attention to the evolution of standards concerning the rights of indigenous populations, taking into account both the similarities and differences in the situations and aspirations of indigenous populations throughout the world." In 1984, the Sub-Commission directed the Working Group to focus its attention on the preparation of standards on the rights of indigenous populations, and accordingly to consider the drafting of a body of principles on indigenous rights based on relevant national legislation, international instruments, and other judicial criteria and consider the situation and aspiration of indigenous populations throughout the world. The working group prepared a draft declaration on indigenous rights. The draft declaration consists of provisions on the protection of intellectual property rights in traditional knowledge. Paragraph 12 of the text provides that "Indigenous people have the right to practice and regenerate their cultural traditions and customs. This incorporates the privilege to maintain, protect and develop the past, present and future manifestations of their cultures, such as archeological and historical sites, artifacts, designs, ceremonies, technologies and visual and performing arts and literature, as well as the right to the restitution of cultural, intellectual, religious and spiritual property taken without their free and informed consent infringing upon their laws, conventions and traditions." Paragraph 29 states that "Indigenous people are entitled to the acknowledgment of the full proprietorship, control and assurance of their social and intellectual property. They have the right to special measures to control, develop and protect their discoveries, technologies and cultural practices, including human and other genetic resources, seeds, medicines, knowledge of the properties of fauna and flora, oral tradition, literatures, designs and visual and performing arts." It has been observed that the TK of indigenous people is not eligible for protection under conventional intellectual property laws and therefore "special attentions" are required. Overall, the Draft Declaration contains provisions that would provide complete attentions of indigenous people and their TK. However, the Declaration is simply a statement of principles with no legal binding status (Donnelly, 1989; Posey and Dutfield, 1996).

4.5 INDIGENOUS AND LOCAL PEOPLE'S CONCERNS IN THE GLOBAL ENVIRONMENTAL AGENDA

Issues of indigenous people and rights of local people have been widely reviewed in global environmental processes. The World Commission on Environment and Development (WCED) established in 1982 by the United Nations General Assembly devoted attention to solve the issues of indigenous people, particularly their knowledge in the sustainable development process. The Commission found that "Tribal and indigenous individuals will require unique consideration as the force of financial advancement disrupt their traditional lifestyles that can offer modern societies many lessons in the management of resources in complex forest, mountain, and dry land ecosystems. Some are undermined with virtual elimination by insensitive improvement over which they have no control. Their customary rights ought to be perceived and they ought to be given a definitive voice in planning strategies about asset advancement in their zones." The Commission calls for "the recognition and protection of their traditional rights to land and other resources that sustain their way of life rights they may define in terms that do not fit into standard legal systems." It additionally suggests that nearby organizations through which indigenous and neighborhood individuals mingle and direct their financial exercises ought to be fortified. Despite the fact that it did not unequivocally address the topic of licensed innovation insurance of customary learning, it made a political system for tending to these issues inside ecological circles. The United Nations Conference on Environment and Development (UNCED) held in 1992 at the recommendation of WCED talked on the issues of IPR in TK and their innovations. Agenda 21 adopted by more than 160 states at the UNCED contains a whole chapter on indigenous people' concerns and makes a wide range of recommendations on how these people' rights should be preserved without further hindrance (Mugabe, 1998; Cullet, 2007).

Agenda 21 (Chapter 26) begins by stating that indigenous people (which account for 5% or 370 millions) of the global population have developed a holistic relationship with the natural environment. Over generations to generations, these people have developed a "holistic traditional scientific knowledge of their lands, natural resources, and environment." It observes that "indigenous individuals and their groups might appreciate the full measure of human rights and fundamental freedoms without obstruction or discrimination" and prescribes that administrations ought to embrace approaches as well as legal instruments that will secure scholarly and social property of indigenous individuals. Another output of the UNCED also recognizes the role of indigenous and local people in global efforts to achieve sustainable development. Principle 22 states that "indigenous people and their communities have a critical role in the management of environment because of their knowledge, skill and traditional practices. States ought to perceive and properly strengthen their character, culture and interests and empower their powerful cooperation in the accomplishment of reasonable advancement." This view is echoed by the Forests Principles, which is also adopted at UNCED. It is stated in Section 5(a) that "the identity, culture and the rights of indigenous people, their communities and other communities and forest residents should be recognized and duly supported through national forest policies. Suitable conditions should be encouraged for these groups to enable them to have an economic stake in forest use, achieve/maintain cultural identity and social organization, as well as adequate levels of livelihood and well-being, through, *inter alia,* those land tenure arrangements which serve as incentives for the sustainable management of forests." Section 12(d) suggested that "welfares appearing from the utilization of indigenous knowledge should therefore be justifiably shared with such people." The CBD, signed by almost 150 states during UNCED, also clearly acknowledges the rights of indigenous and local people in TK and innovations. It also explains about "the traditional dependence of many indigenous and local communities on biological resources, and the need of sharing equitably benefits arising from the use of traditional knowledge, innovations and practices relevant to the conservation of biological diversity and the sustainable use of its components."

Articles 8(j), 10(c), and 18.4 state the rights of indigenous and local people. Article 10(c) provides that each Contracting Party shall protect and encourage customary use of biological resources in accordance with traditional cultural practices that are compatible with conservation or sustainable use requirements. Article 18.4 describes technologies broadly to include "indigenous and traditional technologies." Article 8(j) is perhaps the most authoritative provision dealing with traditional knowledge. It provides that each Contracting Party, as far as possible and as appropriate, "subject to its national legislation, respect, preserve, and maintain knowledge, innovations and practices of indigenous and local communities embodying traditional lifestyles relevant to the conservation and sustainable use of biological diversity and promote their wider application with the approval and involvement of the holders of such knowledge, innovations and practices and encourage the equitable sharing of the benefits arising from the utilization of such knowledge, innovations and practices." Second, Article 8(j) does not talk of protection of the knowledge but merely calls on parties to "respect, preserve and maintain" it. It does not guarantee indigenous and local people any rights in TK (Mugabe, 1998).

Limitations of Article 8(j) have been recognized by parties to the Convention. This is implicit in a number of the decisions that the Conference of Parties (COP) to the Convention has so far made. For example, the third COP held in Argentina in November 1996 agreed (in Decision III/14) on the need to "develop national legislation and corresponding strategies for the implementation of Article 8(j) in consultation with representatives of their indigenous and local communities." The Parties also agreed to establish an inter-sessional process to advance further the work on the implementation of Article 8(j) and related provisions. In support of this process, the Executive Secretary of the CBD was requested by the COP to prepare background documentation on the following issues: (i) consideration of linkages between Article 8(j) and such issues as technology transfer, access, ownership of genetic resources, IPR, alternative systems of knowledge protection, and incentives; (ii) elaboration of key terms of Article 8(j); and (iii) a survey of activities undertaken by relevant organizations and their possible contributions to Article 8(j). Paragraph 9 of Decision III/14 recommended that a workshop on traditional knowledge and biodiversity be convened, prior to the fourth COP, to deliberate on the implementation of Article 8(j), assess priorities for the future work by Parties and by Conference of the Parties, and provide advice to COP on the possibility of developing a work plan on Article 8(j) and related provisions including modalities for such a work plan. In response to this decision, a Workshop on TK and Biological Diversity was held in Madrid, Spain, from November 24 to 28, 1997, at the invitation of the Government of Spain (Mugabe, 1998; Cullet, 2007).

The Madrid workshop discussed a wide range of issues. There was consensus at the workshop that Article 8(j) of the CBD did not provide an adequate legal basis for protecting knowledge and innovations of indigenous people. Several of the participants called for a thorough re-examination and revision of current intellectual property protection systems to create flexibility for protecting indigenous knowledge and innovations. Others called for the establishment of a *sui generis* system that recognizes collective rights of indigenous and local people. It is important to note that some of the participants at the workshop argued that indigenous people are people with inalienable *a priori* rights and therefore they, in these rights, qualify to be parties to the Convention. A document prepared for the fourth COP by the Executive Secretary of the Convention states that many governments are not implementing Article 8(j). None of the studies submitted by governments and other bodies to the CBD Secretariat "refers to a single piece of legislation which specifically addresses the implementation of Article 8(j), but rather, its implementation is carried out, sometimes indirectly, through provisions contained in a wide variety of statutes regarding such matters as land tenure, protected areas, protection of endangered species, land development, water quality, and so on. This wide variety of statutes is sometimes further complicated because similar legislation often exists at national, sub-national, and local levels, with resultant inconsistencies." Concerns on intellectual property protection of traditional knowledge have occupied the agenda of the COPs. The third COP called for dissemination of case studies on the relationships between IPR and the knowledge, innovations, and practices of indigenous and local communities. COP 4, in Decision IV/9, recognized the importance of making IP -related provisions of Article 8(j) and related provisions of the Convention on Biological Diversity and provisions of international agreements relating to IP mutually supportive, and the desirability of undertaking further cooperation and consultation with the World Intellectual Property Organization (Mugabe, 1998; McMains, 2003; Cullet, 2007). On the whole, these efforts are being made a result of the recognition that the Convention does not contain adequate legal obligations to protect any property rights of indigenous and local people in their traditional knowledge.

4.6 TOWARD ALTERNATIVE REGIMES

The previous sections have shown that conventional international intellectual property law could not protect the traditional knowledge of indigenous and local people. The international community has perceived that there is need to devise new administrations or amplify existing ones to oblige the security of traditional knowledge. So far, no intelligent and comprehensive worldwide endeavors are being made to address this worry. There are various choices that nations could endeavor to secure conventional learning of indigenous and nearby individuals. While there is unreasonable consideration being set on licenses and their prohibitive nature in connection to the insurance of conventional information, exchange mysteries have not been sufficiently abused by national establishments and nearby individuals to secure the learning. Indigenous people have used and possibly continue to use trade secrets to protect their information. However, this form of protection of traditional knowledge is generally not institutionalized. It is in this manner critical that national enactment be amplified to contain particular measures that would empower indigenous and neighborhood individuals to apply exchange insider facts to ensure their insight and advancements. Such measures may incorporate unequivocal enunciation of customary information as topic for assurance through prized formulas. Also, there are an extensive variety of institutional boundaries to the commercialization of customary information and advancements in current financial space. For instance, current monetary arrangements of most nations are antagonistic to the immediate utilization of customary developments and situation of such advancements on present day financial space. They fail the test of rigidly established industrial standards. Such strategies ought to be assessed with the perspective of making them more accommodative of conventional information and advancements. There is a need for extensive research to be carried out to explore the potential application of trade secrets. The investment in such studies for their exploration by World Intellectual Property Organization (WIPO) and organizations such as the African Center for Technology Studies (ACTS), the World Conservation Union, and UNEP could add an additional advantage.

Countries should finance in the creation of *sui generis* legislation suitable to their cultural and political conditions. They ought to investigate the advancement of frameworks that will above all else secure conventional learning as protected innovation of indigenous and neighborhood individuals. Such frameworks ought to likewise support (or even require) the stream of advantages from bio-prospecting to indigenous and neighborhood individuals. According to Dutfield, "legislation could be drafted in such a way as to allow a community to become the successor in title of Discovery and Development process. Under this understanding, indigenous groups would have the privilege to ensure customary works on using protected innovation rights systems, halting the typical apportionment by others of the business esteem emerging from their insight. As a correct holder, they would have selective rights to withhold from outsiders them agree to make, utilize, an offer available to be purchased, or import the plant assortment that they created." It is vital that new research be led on customary types of licensed innovation and how traditional knowledge was/is secured by indigenous and neighborhood individuals in various parts of the world. Case studies revealing how indigenous and local people perceive intellect and whether they treat it as property worth protecting would be useful. These works may lead to establish property protection regimes suitable for traditional knowledge and innovations for national and international processes (Dutfield, 1997).

4.7 CONCLUSIONS

This book chapter provides detailed insight on current developments on the protection of TK in the era of emerging IP protection regime. In our view, conventional IP law does not adequately cover or protect TK and innovations of indigenous and local people. However, nonpatent forms of IP protection could be exploited to protect the knowledge and innovations. For example, trade secrets and trademarks offer flexibility for protecting TK and innovations. In addition, Geographical Indications (GI) and PBRs could be used to protect IPR of indigenous people. Indigenous and local people do not have strong institutional arrangements to safeguard their property and enforce trade secrets and trademark in modern economic space. The GI can contribute to development in rural areas. The entitlement to use a GI generally lies with regional producers, and the added value generated by the GI accrues therefore to all such producers. Because GI products tend to generate a premium brand price, they contribute to local employment creation, which ultimately may help to prevent rural exodus. In addition, GI products often have important spin-off effects, for example, in the areas of tourism and gastronomy.

KEYWORDS

- **biodiversity**
- **culture and heritage**
- **indigenous knowledge**
- **IP laws**
- **ITK**

REFERENCES

Agrawal, A., (1995). Indigenous and scientific knowledge: Some critical comments, indigenous *Knowledge and Development Monitor: Development and Change*, *26*(3), 3–4.

Axt, R. J., Lee, M., & Ackerman, D. M., (1993). Biotechnology, indigenous people, and intellectual property rights, *CRS Report for Congress*, 97–478.

Brush, S. B., & Stabinsky, D., (1996). *Valuing Local Knowledge: Indigenous People and Intellectual Property Rights*. Island Press, Covelo, ISBN 1559633786.

Cullet, P., (2007). Human rights and intellectual property protection in the TRIPS era, *Human Rights, 29*, 403–430.

Dutfield, G., (1997). *Can the TRIPS Agreement Protect Biological and Cultural Diversity and Biopolicy International No. 19*, ACTS Press, Nairobi, ISBN 9966410996.

Dutfield, G., (2001). TRIPS-related aspects of traditional knowledge, *Case Western Reserve Journal of International Law*, *33*(2), 233–275.

Farnsworth, N. R., Akerele, O., Bingel, A. S., Soejarto, D. D., & Guo, Z., (1985). Medicinal plants in therapy. *Bulletin of WHO*, *63*, 965–981.

Goldstein, P., (1997). *Selected Statutes and International Agreements on Unfair Competition, Trademark, Copyright and Patent*, The Foundation Press, Inc. New York, 1–420.

Greaves, T., (1994). Intellectual property rights for indigenous people: A sourcebook, *Society for Applied Anthropology*, Oklahoma City, USA, 1–296.

ICSI, (2013). Professional programme, Intellectual property rights-law and practice, The Institute of Company Secretaries of India (ICSI) House, 22, Institutional Area, Lodi Road, New Delhi.

Juma, C., & Ojwang, J. B., (1996). In: *Land We Trust: Environment, Private Property and Constitutional Change,* ACTS Press, ISBN 9966-42-042-8.

Juma, C., (1989). *The Gene Hunters: Biotechnology and the Scramble for Seeds*. Zed Books and Princeton University Press, London and Princeton. ISBN 0-86232.

Lewis, W. H., Elvin-Lewis, M. P., & Fast, D. W., (1991). *Pentagonia gigantifolia* (Rubiaceae) as a Snakebite Remedy: Empirical methodology functioning in Amazonian traditional medicine. *Economic Botany*, *45*, 137–138.

Mashelkar, R. A., (2001). Intellectual property rights and the third world. *Current Science*, *81*(8), 25.

McManis, C. R., (2003). Intellectual property, genetic resources and traditional knowledge protection: Thinking globally, acting locally, *Cardozo Journal of International Comparative Law*, *11*(2), 547–583.

Moran, K., King, S. R., & Carlson, T. J., (2001). Biodiversity prospecting: lessons and prospects. *Annual Review of Anthropology*, *30*, 505–526.

Mugabe, J., (1994). "*Technological Capability for Environmental Management: The Case of Biodiversity Conservation in Kenya*" (PhD, Dissertation Submitted to the University of Amsterdam, The Netherlands).

Mugabe, J., (1998). Intellectual property protection and traditional knowledge. In: *Intellectual Property and Human Rights*, 50th Anniversary of the universal declaration of human rights. World Intellectual Property Organization, 97–123.

Posey, D., & Dutfield, G., (1996). *Beyond Intellectual Property*, International Development Research Center, Ottawa, ISBN-13: 978-0889367999.

Posey, D., (1991). Intellectual property rights for native peoples: challenges to science, business, and international law. In: *International Symposium on Property Rights, Biotechnology and Genetic Resources, Nairobi Kenya.*

Posey, D., (1994). In: Sanchez, V., & Juma, C., *Biodiplomacy: Genetic Resources and International Relations*, African Center for Technology Studies (ACTS) Press, Nairobi.

Quinn, M. L., (2001). Protection for indigenous knowledge: An international law analysis, *St. Thomas Law Review*, *14*, 287–313.

RAFI, (1994). *Conserving Indigenous Knowledge: Integrating Two Systems of Innovation*, A study prepared for the United Nations development program (UNDP), New York, 48–54.

Schuster, B. G., (2001). A new integrated program for natural product development and the value of an ethnomedical approach, *Journal of Alternative and Complementary Medicine*, *7*, 61–72.

CHAPTER 5

PROTECTING PLANT BREEDERS' RIGHTS IN AGRICULTURAL RESEARCH AND DEVELOPMENT

HILLARY MIREKU BORTEY and STEPHEN AMOAH

Council for Scientific and Industrial Research – Crops Research Institute, P.O. Box-3785, Kumasi Ghana, E-mail: hmireku@gmail.com

CONTENTS

ABSTRACT

The advent of TRIPS agreement under WTO led to a drastic change in the intellectual property rights (IPR) in agricultural research and development. The International Union for the Protection of Plant Varieties (UPOV) came in 1961 with the provisions of plant breeders' right and protection of plant varieties. However, many countries did not adopt the provisions made under UPOV 1961. Food security is now a global issue particularly in developing countries that is critically important to meet the immediate demand of food. Food production achieved by development of new plant varieties is always encouraging. However, a path-breaking achievement in food production was made by green revolution. Presently, enactment of several national and international regulations on germplasm exchange is restricting free exchange of germplasm. Plant breeders' rights (PBRs) are encouraging breeders for technology development (mostly seed) and their exchange, thus earning money and strengthening economic status of the organization. The PBRs are affecting both developed and developing countries directly and indirectly.

5.1 THE ROLE OF AGRICULTURAL RESEARCH IN MEETING INCREASING FOOD DEMAND

It is estimated that if global population reaches 9.1 billion by 2050, world food production will need to rise by 70% and food production particularly in the developing world will need to double (FAO, 2009). What is of critical importance about this projection is that increase in food production will have to overcome rising energy prices, the continuing loss of farmland to urbanization, and increased drought and flooding resulting from climate change (FAO, 2009). Agricultural research has been and continues to be a major factor contributing to agricultural growth and economic development all over the world. This contribution of agricultural research has led to increase in food production and lowered food prices (MaCalla and Brown, 2000).

Unfortunately, despite the many successes achieved in the past by agricultural research and development (R&D), many developing countries continue to face the issues of food insecurity, malnutrition, and poverty. Many of these countries especially in Africa are projected to be "hot spot" for hunger and malnutrition for many years to come (Rosegrant, 2001). The rhetoric has been how these countries can alleviate poverty, reduce malnutrition, and improve food security under the current myriads of challenges posed by climate change and other socio-economic dynamics in the world.

According to Maredia (2001), an infusion of new technologies that can transform an economy from subsistence agriculture to a more productive commercialized system is an essential component of the solution to improve food security and reduce malnutrition and poverty. Therefore, the need to consciously invest in agricultural research, particularly plant breeding, to improve crop productivity and development of modern agricultural technologies is timely and imperative.

5.2 CONTRIBUTION OF PLANT BREEDERS' RIGHTS TO AGRICULTURAL GROWTH AND DEVELOPMENT AND FOOD SECURITY

Intellectual property (IP) is a term that has been used to refer to the general area of law that encompasses patents, trademarks, designs, and a host of other related rights (Bently and Sherman, 2009) including plant breeders rights (PBRs). IP law creates property rights in wide and diverse range of intangible things. Kameri-Mbote et al. (2013) also puts intellectual property rights (IPR) as bundle of rights and expectations in a tangible or intangible thing that are enforceable against third parties including the government. All IPRs are negative in nature as they are primarily rights to stop others from doing certain thing(s) with regard to that what is protected. Such rights include the right to stop pirates, counterfeiters, imitators, and even third parties from using what is protected without the license or permission from the rights owner (Cornish et al., 2013). IPRs are granted by the State, and they allow the owner to exercise some level of State power against third parties (Merges, 2011). The owner of the IP asset is able to exercise these rights after the State recognizes the existence of the invention and protects it by allowing him to exercise the State power against third parties (Muchiri, 2014).

The significance of IP generally in promoting economic development has been widely "preached," although it has been divergent. Some are of the view that IP regimes seek to encourage the protection of investment in research and development (R&D) by industry, thereby spurring innovation (Repetto and Cavalcanti, 2009; Kanwar and Evenson, 2001). The current IP system is based on the economic theory stemming from the utilitarian justification as influenced by Bentham. Under this theory, incentives are provided to encourage creators to engage in innovative activities while at the same time being rewarded for their creativity (Wekesa, 2005). Others argue that the economic justification lies in the former goal of providing incentives rather than in the latter goal of rewarding creativity (Merges et al., 2003). The consequence of this is that IPRs limits the diffusion of knowledge, thereby preventing those without the creator's authority from using them. On the contrary, Cullet (2003) and Jordens and Button (2011) are of the view that strong IPRs play a significant role in attracting investment in agriculture and enhance market growth, access, and diversification as they provide incentives to breeders by assuring that their expenditure and innovations or creativity will be protected In a global knowledge-based economy, IPRs are critical to the international competitiveness of both nations and firms (Langford, 1997). Thus, adequate IP protection at an international scale has become essential for appropriating global revenue streams to support investment in developing state-of-the-art technology (Maredia, 2001) by both public and private sectors, especially in agricultural R&D. The above justification notwithstanding, there seem to be some level of skepticism regarding the role of IPR in general and PBR in particular in ensuring food security.

5.3 DOES IPR HAVE ANY RELATIONSHIP WITH FOOD SECURITY?

There is definitely some existence of linkages between food security as a goal in terms of policy, economic, and legal and IPRs as an instrument to promote and enhance human creativity and overall social well-being (Narasmhan et al., 2011). Food security is part of the basic human right to food and broadly defined as timely access to sufficient and nutritious food (Barton, 2006). Food security is linked to IP as much as plant variety protection (PVP) or PBR, and patents, as applied to genetic resources, biodiversity components, and biotechnological processes. This is because of the tendencies of IPR to limit the possibilities of farmers to freely grow certain crops, and for people to consume resulting agricultural products (Narasmhan et al., 2011). The latest report by FAO, IFAD, and WFP (2014) estimates that 805 million people are chronically undernourished (1 out of every 9 persons) globally over the period of 2012–2014. The vast majority of these undernourished people live in developing countries.

In Africa, there has been insufficient progress toward international hunger targets, especially in the sub-Saharan region, where more than one in four people remain undernourished—the highest prevalence of any region in the world (FAO, IFAD, and WFP, 2014). The Rome Declaration, which was issued by the Summit, pledged to reduce the number of hungry people by half by 2015. This goal was also included in the Millennium Declaration of the United Nations in 2000. This objective required the number of undernourished to fall at a rate of 20 million per year. However, data in 2001 indicated that the rate of decline was less than 8 million per year (FAO, 2001). The plan of action adopted at the World Food Summit (WFS) identified poverty eradication as a prerequisite for the attainment of food security and stressed the importance of access to land, water, improved seeds, and plants; appropriate technologies; and farm credits to achieve this goal (FAO, 2001).

Historically, the periods between 1960 and 1990 witnessed a massive increase in food productivity, which was described as the Green Revolution, achieved by increasing the productivity of cereals, expanding the area of arable land, and by massive increase in the use of fertilizers and insecticides. On the contrary, to meet the food security needs of the next

30 years and create wealth in poor communities, there is a need to increase agricultural productivity on the presently available land, while conserving the natural resource base (Conway, 1999). Thus, reliance upon the chemically nurtured, high-yielding crop varieties of the past is no longer economically or environmentally acceptable. A second Green Revolution is required that combines traditional agronomic wisdom with modern agricultural science (Serageldin and Persley, 2000). It is in this area of agricultural innovation that IPR plays an important role.

One of the most important mechanism currently available and been used widely for legally protecting agricultural innovations is PVPs or PBRs (Blackeney et al., 1999). It must, however, be noted that several international frameworks exist for the protection of plant materials both for food consumption and for other important uses such as cosmetics and pharmaceuticals among others.

PVPs/PBRs have been proven to encourage investment in plant breeding and open a country's door to overseas varieties where the protection of law is guaranteed (Jordens and Button, 2011). More often than not, new plant varieties render higher yields and quality products, as well as greater resistance to disease, rendering them a crucial aspect of production. Through these varieties, the benefits of plant variety protection therefore have the tendency of improving food security. The IPRs, generally, and PVPs, particularly, are deemed to provide assurance to breeders to recover invested capital and cost of a value-added innovation (Lesser, 1997; Singh, 2007; Dutfield, 2011). This will stimulate technology transfer and further development. Thus, the basic principle behind the legal protection to plant varieties is to encourage commercial plant breeders to invest in agricultural R&D for breeding new plant varieties and improve existing plant varieties (Lesser, 1997) for increased crop productivity.

Moreover, developing countries' dependence on foreign technology suppliers, especially improved foreign varieties of crops, which often are not locally adaptable, is likely to reduce. It is further recognized that an effective PVP or PBR system can provide important benefits in an international context by removing barriers to trade in varieties, thereby increasing domestic and international market scope (Jordens and Button, 2011). Other potential role of PVP in ensuring food security is the relative access to foreign-bred varieties to augment locally developed materials. Accordingly, as argued from the economic theory perspective, the PVP system will provide incentives to encourage breeders to engage in innovative activities while at the same time being rewarded for their creativity.

5.4 DIVERGENT VIEWS ON THE CONTRIBUTION OF IPR/PBR ON FOOD SECURITY

On the contrary, another school of thought perceives IP regimes a restriction to inventor's ability to obtain and exploit foreign sophisticated technology (Kanwar and Evenson, 2001). From a food security perspective, some are of the view that the economic theorem justification for protecting plant varieties has grave consequences chiefly in a system already fraught with problems as exist in most agricultural sectors of developing countries. In respect of PVP systems, Barton (1998) earlier had suggested it is "doubtful" whether developing countries should enact PBRs because the "trade-offs are quite different" compared to developed countries. This was corroborated by Rangnekar (2002) that it is uncertain as to whether the availability of PVP/PBR system results in increase of varietal release, as well as whether it is an economically good as claimed.

Further, a report by International Plant Genetic Resources Institute (IPGRI, 1997) indicated that developing countries whose economy are mainly dependent on traditional agriculture, which is largely characterized by smallholder farmers, have less to gain from strong PVP systems. Of course, the report does not suggest the absence of a PVP system altogether, but rather the degree of protection must be put in the right context. Further, these authors are of the view that a PVP regime that makes no provision for farmers' rights implies discrimination against farmers compared to commercial breeders by omitting means of recognition and compensation for efforts and contributions by farmers and indigenous communities (Singh, 2007). Singh (2007) argues one important feature of subsistence farming is that the traditional varieties grown by subsistence farmers contain a lot of genetic diversity. These genetic resources provide the foundation upon which plant breeding depends for the creation of new varieties, thereby contributing to food security (Singh, 2007) and should be recognized accordingly. Most critiques are of the view that PVP systems or PBRs restricts farmers' seed saving tradition, and this has direct food security consequences (Wekesa et al., 2009). In the same argument, any act, regulatory or otherwise, that results in higher seed prices jeopardizes the situation given that lack of sufficient financial resources to buy food is a major cause of individual food insecurity (Oxford: Oxfam, 2002). Singh (2007) further claimed that a PVP system increases the rate of varietal release, which results from increased private sector involvement in plant breeding. However, this is accompanied by a shortening life-span of varieties, which affects farmers' decision-making on adoption of improved varieties.

Donnenwirth et al. (2004) is of the view that an effective IP or PVP regime should not only encourage access to varieties that are already well-adapted and high performing use in breeding as this may reduce long-term on-farm performance gains; increase vulnerabilities to pests and diseases; and reduce the diversity of germplasm. Rather, the PVP system should encourage the development and use of broad crop categories in the cultivation chain.

5.5 IPR: PLANT BREEDERS' RIGHTS AS "BAIT" FOR ATTRACTING PRIVATE SECTOR INVESTMENT IN AGRICULTURAL RESEARCH AND DEVELOPMENT

Agricultural research in developing countries is predominantly the domain of public institutions (Maredia, 2001; Bortey and Mpanju, 2016). These institutions mainly comprising National Agricultural Research Institutes (NARIs) and Agricultural Universities have been supported financially and technically by the international donor funding agencies and partners. Generally, it is known that public investment in the development of genetically improved varieties can increase agricultural productivity and continues to be a prerequisite for helping lift millions out of poverty and banishing hunger and malnourishment (Donnenwirth et al., 2004). It is thus not surprising that decades after plant breeding emergence was recognized as a field of science in the late 19th century, almost all plant breeding activities took place in the public institutes (Morris et al., 2006). This phenomenon could be attributed to several major trends that are transforming global agriculture and in the process altering the environment in which agricultural research organizations operate (Morris et al., 2006). Consequently, these changes have affected the way plant breeding research gets funded and carried out.

For instance, as most industrialized countries' agriculture becomes more and more commercialized, private firms tend to increase their investment in research, and thus, the argument in favor of using public funds to support agricultural research becomes increasingly difficult to sustain (Morris et al., 2006). On the contrary, in many developing countries, agriculture tends to be more subsistence-oriented and small-scaled, thus less attractive to the private sector investments; hence, the arguments in favor of public investments in agricultural research are much stronger.

In the early 1980s and 1990s, a study by Tabor (1998) indicated that on average, developing countries allocated 4 to 10 times more of their limited resources on agricultural research than the higher-income Organization for Economic Co-operation and Development (OECD) countries. As a share of gross domestic product, Africa allocated 0.3% of GDP to agricultural research, Asia 0.1%, and Latin America 0.05% (Tabor, 1998). In the early 1990s, OECD governments allocated 0.17% of public funds to agricultural research compared with 0.23% in Latin America, 0.6% in Asia, and 0.7% in Africa (Tabor, 1998). Within the same period, Pardey et al. (1997) observed that despite the fact that agricultural research was receiving higher priority in government budgets, the rate of growth in public investment declined sharply. In developing countries, the growth rate of public investment in agricultural research fell from 6.4% in the 1970s to 3.9% in the 1980s.

Despite the contributions, agriculture and agricultural research offer to economic development in most developing countries, public investments in agricultural research have not kept pace with these acknowledged needs, particularly in sub-Saharan Africa and other developing countries. In terms of agricultural R&D spending, Pardey et al. (2006) stated that of all types of agricultural expenditures, spending on agricultural R&D is the most crucial to growth in agriculture. In 2000, developing countries spent 0.5% of agricultural GDP on R&D. In the same year, developed countries as group spent 2.4% of agricultural GDP on R&D (Pardey et al., 2006). The World Bank study by Akroyd and Smith (2007) on some selected countries showed a sharp decline for total agriculture share to total expenditure as a percentage of GDP between 1980 and 2002 (Figure 5.1).

A more recent study by Bortey and Mpanju (2016) on the possible implication of implementing a PBR system in Ghana indicated that the major funding for the development and release of new crop varieties for

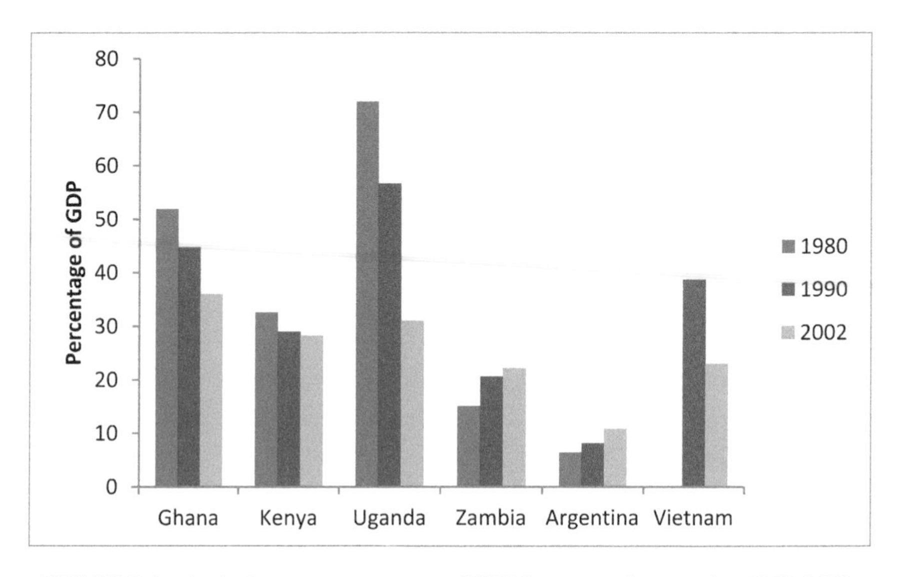

FIGURE 5.1 Agriculture as a percentage of GDP in case study countries, 1980–2002.

the past 10 years was mainly from external donor support. Unlike the developed countries where the private sector amounts to almost half of total agricultural research expenditures (Pardey et al., 1997), the story cannot be told of the developing world. The increasing share of the private sector investment in agricultural research in the developed countries according to Maredia (2001) could partly be attributed to the expanded IPR over biological innovations. With this trend of public expenditure on agricultural R&D, global food securities is increasingly becoming dependent upon research and product development by the private sector.

Ironically, the private sector's efforts to develop new varieties of plants and other agricultural-related technologies are heavily influenced by effective IP regimes. The existence of such IPR determines the level of risk they have to take, and thus, the kinds of research that can be profitably pursued (Donnenwirth et al., 2004).

For instance, Pardey et al. (1997) observed that in five OECD countries, where about 80% R&D is by the public, with 12% being involved in private research, the research of these private sector players are focused on post-harvest technologies and chemical research on fertilizers, herbicides, and pesticides.

Countries that have full obligation under the WTO/Trade and Related Aspects of Intellectual Property (TRIPS) Agreement are required to provide protection for all types of plant materials have either done so through patents or by a form of *sui generis* system based on PVP or PBRs (Louwaars et al., 2005). As a result, these countries have provided incentives for private firms to invest in plant breeding research (Morris et al., 2006) and seen rapid growth in the number of private seed companies and greater availability of improved crop varieties (Morris and Ekasingh, 2002). In spite of the perceived drawbacks and uncertainties regarding the contribution of PBR in encouraging private sector investment in agricultural R&D, evidence so far shows that it is the most effective system available. Countries and governments are to evaluate the various PVP systems and make policy decisions based on their individual peculiarities.

5.6 PLANT BREEDING, VARIETAL DEVELOPMENT, AND PLANT BREEDERS' RIGHT

New plant varieties represent an output of plant breeding, which is the process of generating, developing, or producing new plant varieties with better or new desirable features (Watal, 2001). Plant breeding is fundamental in agricultural research as it creates plant varieties with desirable qualities such as increased yield of crops, high resistance to pests and diseases, and harsh climatic conditions among others (Louwaars, 2009). However, the breeding process itself is lengthy and often costly, while the process of duplication of the new varieties is easy. In developing new plant varieties, new techniques are employed, ranging from conventional breeding (crossing and selection), tissue and cell culture techniques, and molecular biology breeding to genetic engineering. The newly generated or developed variety using the above techniques may have a reproductive material that makes them distinctive from each other. The process resulting from the creation of the new variety involves human intellect, skill, and art.

Thus, it is morally justified to grant protection to the new variety to enable the breeder "own" and have the "exploitative" right. Plant breeding and the new variety are recognized and protected through a formal process of granting exclusive rights to the plant breeder. Although there exist other forms by which plant breeders can protect their newly developed variety, PVP/PBR has widely been regarded as the most suitable form of protection. It is considered to be an equitable way to give plant breeders an opportunity of a fair reward for their work, effort, and investment in breeding (Bently and Sherman, 2009). The right protected under a new variety extends to the production, reproduction, and conditioning for the purpose of propagation, offering for sale or any other marketing activity, exporting or importing and stocking of the propagating material (UPOV 1991 Act, Art. 14). Protection of new plant varieties in whichever form is mainly justified by the argument that it is an incentive for R&D. The World Intellectual Property Organization (WIPO) justifies the granting of IPR to inventions as a way of giving expression to the moral sentiment, that a creator should enjoy the fruits of their creativity and to encourage investment in skill, time, finance, and other resources into innovation in a way that is beneficial to the society (WIPO, 2004).

The implementation of the PBR will provide access to a critical input to agricultural productivity, which is seed. This system will provide incentives for the technological advancement necessary for economic growth and development, facilitate technology transfer and access to foreign varieties, stimulate investment, including that of foreign firms, and encourage local breeders and seed industry (FAO, 2000). The PBRs are intended to provide the breeders with an opportunity to recoup their profits while providing the growers with better and improved plant varieties (Dutfield, 2011). The grant affords them some protection against unauthorized duplication of their varieties (Bently and Sherman, 2009). Such protection is also considered as an incentive to the breeder to undertake more research and develop more varieties to the benefit of the stakeholders. Under the UPOV system, PBRs for all crops shall be protected over a minimum period of 20 years and 25 years for trees and vines (UPOV, 1991; Article 19).

5.7 INTERNATIONAL LEGAL FRAMEWORKS FOR THE PROTECTION OF PLANT VARIETIES AND BIOLOGICAL RESOURCES

Despite the existence of certain international provisions for protection of inventions and creativity in the developed world in the 1980s, aspects related specifically to the protection of plant varieties and biological resources has been recent. In particular, there has been increased attention in the past few years to strengthening IPRs in plant breeding research in developing countries. One of such international agreements that sought to push this agenda is the TRIPS Agreement (TRIPS, 1994) of the World Trade Organization (WTO). This agreement requires all WTO members to introduce at least a minimum level of protection in their national laws for plant varieties and inventions in biotechnology (Article 27.3(b) of TRIPS, 1994). It must, however, be noted that the TRIPS agreement is not the only international agreement related to regulatory systems for plant variety protection. The International Treaty on Plant Genetic Resources for Food and Agriculture (ITPGRFA), the Convention on Biological Diversity (CBD), The International Union for the Protection of Plant Varieties (UPOV) systems and the Intergovernmental Committee on Intellectual Property and Genetic Resources, Traditional Knowledge, and Folklore of the World Intellectual Property Organization (IGC Text, WIPO), which is currently under discussions are all regimes for protection of plant varieties. Ghana is signatory to all but the latter, which is in draft stage.

5.7.1 THE CONVENTION ON BIOLOGICAL DIVERSITY (CBD)

The 1970s and 1980s saw a number of initiatives to halt the loss of species and the destruction of habitats and ecosystems. A consensus gradually emerged that the Earth's priceless reservoir of biological diversity could be saved only through international cooperation and funding, based on the introduction of a suitable international legally binding instrument. Consequently, the Convention on Biological Diversity (CBD) was adopted at the Rio de Janeiro Earth Summit, in June 1992 (Repetto and Cavalcanti, 2009). Over 150 States signed the documents at the Rio conference, and since then more than 196 countries are parties to the Convention (https://www.cbd.int/information/parties.shtml).

The Convention is a legally binding framework agreement. Thus, Article 6 states, each Contracting Party is required to, "… in accordance with its particular conditions and capabilities, develop national strategies, plans or programmes for the conservation and sustainable use of biological diversity."

The main objective of the Convention is the conservation of biological diversity, sustainable use of its components and fair and equitable-sharing the benefits arising from the utilization of generic resources (Article 1). The CBD is the most recent and comprehensive instrument that balances the need for legal protection of IPRs with that of the specific context of developing countries sustainable development agendas.

The CBD recognizes both the dependence of local communities on biological resources and the role that they have played in the evolution, conservation, and sustainability of such resources. The CBD is not limited to plants but to all biological resources. It offers an important setting for information sharing but also resources for technical assistance that countries, especially in the developing economies can explore to their benefit. The framework recognizes member state's sovereignty over their biological resources but qualifies this control or sovereignty with the introduction of the notion of "common concern" (Articles 3 and 15.1 of Biological Convention, supra n.2). This implies that the protection of a country's biodiversity is of interest not only to that country but the international community as well. It is, however, worthy to note that Article 16 of the Convention clearly indicates that intellectual property rights are not to undermine the working of the convention.

Therefore, governments, particularly in developing countries, should develop measures and arrangements to implement the rights of countries of origin of genetic resources or countries providing them, as defined in the CBD. Both farmers' and community's rights can be well protected if such measures are taken without necessarily having it enshrined in the PBR Act or other PVP systems.

5.7.2 INTERNATIONAL TREATY ON PLANT GENETIC RESOURCES FOR FOOD AND AGRICULTURE (ITPGRFA)

The International Treaty is a binding treaty and emphasizes the conservation of biodiversity and access and benefit sharing, which equally are the main tenets of the CBD. The Treaty's objectives are "the conservation and sustainable use of plant genetic resources for food and agriculture," and the "fair and equitable sharing of the benefits arising out of their use, in harmony with the Convention on Biological Diversity, for sustainable agriculture and food security" (Article 1.1 of ITPGRFA). The Treaty creates a multilateral system to facilitate access and benefit-sharing. Because it is recognized that each State has sovereign rights over their plant genetic resources, it was necessary that an agreement be reached to pool all these resources together for common good. The multilateral system constitutes a special regime in the context of the international principles of access to genetic resources and benefit-sharing dictated by the CBD. This system is meant to ensure effectiveness, efficiency, and transparency in accessing plant genetic resources and further ensure fairness in the distribution of benefits that may arise in the utilization of the genetic resource. The multilateral system is a global gene pool of crops and forages established by state governments in a binding agreement of international law. The multilateral system pools samples of genetic material from a set of crops, which are listed in Annex 1 to the ITPGRFA. These crops provide about 80% of our food from plants (www.planttreaty.org/mls_en.htm). Samples are included in the gene pool by the state governments and the institutions that they control. Samples also come into the gene pool from international institutions as well as from natural and legal persons—anyone, which is—within the jurisdiction of the contracting parties. Exchange of plant genetic resources (PGRs) through this system is by Material Transfer Agreements (MTAs) and reflects the aim of ensuring food security by facilitating access to breeding genetic material.

The International Treaty is unique in that it provides a direct link between biodiversity conservation, sustainable agriculture, IPRs, and food security. Addressing mutually agreed facilitated access to PGRs, the International Treaty states that "recipients shall not claim any intellectual property or other rights that limit the facilitated access to the plant genetic resources… or their genetic parts or components in the form received from the Multilateral System,"(Article 12.3(d) of ITPGRFA) and that "access to plant genetic resources for food and agriculture protected by intellectual and other property rights shall be consistent with relevant international agreements, and with relevant national laws."(Article 12.3(e), ITPGRFA).

The rights of communities are expressly recognized in the ITPGRFA, which in Article 5 provides that "all efforts on plant improvement should be done in collaboration with the farmer." In addition, the right of farmers to save, use, exchange, and sell farm-saved seeds and propagating material is recognized. However, the Treaty is silent with regard to farmers' rights over their landraces (Cullet, 2003 supra note 119, p. 2). In this regard, the only rights that the International Treaty acknowledges are farmers' residual rights to save, use, exchange, and sell farm-saved seeds.

In other words, although farmers are not granted any exclusive right over their varieties, the Treaty found a way to provide a counterbalance to IPR by establishing, where appropriate, benefit-sharing arrangements consonant with notions of communal, as opposed to individual or private property. Moreover, Article 9(2) states that the responsibility for realizing farmers' rights rests with national governments.

5.7.3 INTERNATIONAL UNION FOR THE PROTECTION OF NEW VARIETIES OF PLANTS, (UPOV) SYSTEM: POTENTIAL IMPACT ON FOOD AND SEED SECURITY IN DEVELOPING COUNTRIES

There has been vigorous debate on the sorts of *sui generis* systems that might comply with Article 27.3 (b) of the TRIPs agreement. Although the TRIPs provision makes no reference to UPOV, which is considered to provide some leeway in the formulation of *sui generis* systems, the plant variety protection system that has become the most important mechanism currently available and been used widely for legally protecting agricultural innovations (Blackeney et al., 1999) is fashioned around the UPOV system.

The International Union for the Protection of New Varieties of Plants (UPOV) pioneered the establishment of a PVP system to provide a legal framework that encourages plant breeding, thereby, responding to the challenges of a changing world (Jordens and Button, 2011). The main aim of UPOV is to encourage the development of new varieties of plants, for the benefit of society through the grant of protection, which serves as an incentive to those who engage in commercial plant breeding. The convention was first adopted in Paris in 1961 and entered into force in 1968. It has been revised three times in 1972, 1978, and 1991.

Majority of countries, particularly those from the sub-Saharan Africa, have their variety protection system (PBRs) reflecting the provisions under the UPOV 1991 Act. Unfortunately, most authors are of the view that the

1991 Act possesses antagonistic provisions in relation to its possible negative implication to food, seed security, and sovereignty. From the perspective of farmers, probably the most contentious aspect of the 1991 Act is the limitation of the farmers' privilege to save seed for propagating "on their own holdings" the product of the harvest which they obtained by planting a protected variety "on their own holdings," "within reasonable limits and subject to the safeguarding of the legitimate interests of the breeder" (Article 15(2) of UPOV, 1991 Act). Unlike the 1978 Act, the 1991 version of the farmers' privilege does not authorize farmers to sell or exchange seeds with other farmers for propagating purposes (Leskien and Flitner, 1997). This provision, many think, is at variance with the practices of farmers in many developing nations, where seeds are exchanged for purposes of crop and variety rotation (Leskien and Flitner, 1997). A recent study on the impact of UPOV-91 Act by the Berne Declaration (2014) revealed that the 1991 Act restrictions on the use, exchange, and sale of farm-saved PVP seeds will make it harder for resource-poor farmers to access improved seeds. According to the same study, this could negatively impact on the functioning of the informal seed system. From a human rights perspective, they are of the view that restrictions on the use, exchange, and sale of protected seeds could adversely affect the right to food, as seeds might become either more costly or harder to access (Berne Declaration, 2014).

On the contrary, an earlier study conducted by UPOV on the impact of the system on some selected countries showed positive effect in some critical areas of the agricultural system, although as indicated the results should be seen in the context of individual situations (Jordens and Button, 2011). For instance, the study revealed that the introduction of the UPOV system was associated with increased breeding activities and the encouragement of new types of breeders including private breeders, researchers, and farmer-breeders. It also led to increased development of partnership including public–private partnership (Jordens and Button, 2011). The study also revealed that the introduction of the UPOV system was associated with increased development and adoption of new improved crop varieties that provided improvements for farmers, growers, industry, and consumers (Jordens and Button, 2011).

The UPOV currently as on April 2016, has a total membership of 74 States. In Africa, there are four members (South Africa, Kenya, Morocco and Tunisia). Zimbabwe application is under consideration and the latest to join is a collective application of 16 West African states under Africa Intellectual Property Organization (OAPI).

5.8 IMPACT OF PLANT BREEDERS' RIGHTS IN DEVELOPED COUNTRIES

There have been several studies that have attempted to analyze the effects of IPR regimes on plant breeding in industrialized countries (Trips et al., 2007). Most researchers to capture the impact of PBRs have studied some variables including level of investment in plant breeding activities in respect of number of new firms and R&D expenditures, new varieties released, and market concentration (Murphy, 1980). A primary rationalization supporting IPRs is the incentive it provides to private investment, a point empirically validated by evidence establishing the increased level of private investment (Rangnekar, 2000). This is argued in respect of the number of new firms becoming active in plant breeding, and a related indicator would be the initiation of breeding programs (Rangnekar, 2000).

Interestingly, the results of these studies are divergent in terms perceived impact of PVPs. In the developed world, a study by Penna (1994) showed an increase in breeding investments in the UK for some horticultural crops, but not other crops, and Diez (2002) demonstrated that there have been more PVP certificates issued in Spain to private sector breeders for crops for which PVP protection was available earlier. An assessment of the effects of PVP in Canada found increased breeding investment for some horticultural crops, but less for field crops (Canadian Food Inspection Agency, 2001). For wheat, Alston and Venner (2002) showed that private sector investment in US wheat breeding has remained static while that of the public sector has increased. Murphy (1980) also reported that private investment in plant breeding in the UK increased by as much as 500% during 1975–1980 with the introduction of legal framework to protect plant varieties.

Furthermore, the World Bank survey on IPRs reviewed a US-based studies and concluded that protection for intellectual property in seeds and plants in developed countries has resulted in increased private breeding activity (Lesser, 1990). For instance, many countries that are members of UPOV have reported increases in plant breeding activities with direct effects upon their agricultural and horticultural industries (UPOV, 2005). These countries also reported increases in the range of varieties made available to farmers and growers as well as increased investments in agriculture.

In Argentina for example, the number of PVP grants to foreign breeders, increased following the amendment of the national PVP law to comply with UPOV. The number of granted titles more than tripled (from 17 to 62) in 10 years since the country became a UPOV member. The increase in number of foreign titles was more evident in important agricultural crops (such as soybean, roses, strawberry, and Lucerne), for which improved varieties are important for competitiveness in the global market.

In China, farmers have greatly benefited from the introduction of a PVP system and in particular with UPOV membership. They have seen the development of a number of new varieties of the most important agricultural crops such as maize, rice, wheat, soybean, and oilseed rape. The number of PVP grants in China increased from 39 to 261, 1 year after joining UPOV. These examples demonstrate that with suitable PVP laws, returns from breeding activities can be potentially reinvested in crop improvement activities.

However, Butler (1996) and Lesser (1997) are of the view that although several studies document an increase in private sector breeding for a number of non-hybrid crops in the USA since the PVP Act of 1970, but most attribute only a modest role to PVP for these changes. For instance, in the Netherlands, Leenders (1976) reported the availability of PBRs was followed by consolidation within the industry that led to disappearance of a number of smaller cereal breeders. In other words, the adoption and implementation of PVP are not the only factor to increasing private sector investment in breeding activities in all jurisdictions, but it plays a critical role in this process.

5.9 IMPACT OF PBRs OR PVP SYSTEM IN DEVELOPING COUNTRIES

Unlike the developed countries, where PVP systems have existed for decades and consequently made impact assessment relatively easy, it has become relatively difficult to do any type of quantitative study on the effects of IPR regimes on plant breeding in developing countries. In Africa, countries who are members of UPOV include Kenya, South Africa, Morocco, and Tunisia and latest to join is OAPI, which comprise 17 countries that are mostly French-speaking as on June 2014. The discussion on impact of PBR in developing countries shall focus on Kenya and South Africa. Further, the focus shall be on R&D, development of new crop varieties, and on the economy as influenced by the adoption and implementation of PBR.

5.9.1 IMPACT OF PBRs IN DEVELOPING COUNTRIES: THE CASE OF KENYA

According to Rangnekar (2006), economic literature to review the impact of PBRs' with respect to R&D is to focus on proxy indicators like the number of applications or grants for PBRs. In Kenya, a number of impact studies (GRAIN, 1999; Cullet, 2001; UPOV, 2005; Sikinyi, 2009) have been conducted. This could be attributed to the fact that Kenya was the second country in Africa to join UPOV in 1999 after operating a *sui generis* PVP system for 2 years. The PVP scheme started to operate in 1997 and Kenya acceded to the 1978 Act of the UPOV Convention in 1999.

A study by GRAIN (1999) reported that the introduction of PVP did not substantially foster the development of new food crops in Kenya and Zimbabwe. The study further asserted that of 136 applications filed and tested since 1997, only one was a food crop, while the majority of applications were for cash crops and ornamental plants. The GRAIN (1999) report relied on the number of applications to make its observations and conclusions, which the author of this review holds a divergent view and believes the conclusions by the GRAIN report could not be entirely accurate. First, any attempt to do a quantitative study on the effects of IPR regimes on plant breeding in developing countries, whose PVP systems are at the infantile stage will be relatively difficult. The trend of application in Kenya changed over time (Sikinyi, 2009). The GRAIN report (1999) was observed just a year after the PBRs system has been formally introduced in Kenya.

The number of residents/nationals who applied for PBR increased gradually and was highest in 2001 (164) out of a total of 197 applications (Sikinyi, 2009).

Regarding the number of new crop varieties developed as a result of the introduction of PBRs in Kenya, Rangnekar (2006) reported that a total of 672 applications were made between 1997 and 2004 (Table 5.1). Majority of the applications from nationals/residents were for food and industrial crops, while foreign applications were for ornamental plants. From Table 5.1, it can be inferred that the number of crops that can secure food security to the people of Kenya constitutes a higher percentage. For instance, from 1997 to 2004, as high as 55 maize, 30 wheat, and 7 each in sorghum and barley varieties have been applied for protection under the system. According to Wekesa et al. (2003) and Schroeder et al. (2013), maize is the primary staple crop in Kenya and plays an important role in the livelihood of the people of Kenya. Its availability and abundance determine the level of welfare and food security in the country (Schroeder et al., 2013).

It is further estimated maize accounts for more than 20% of total agricultural production and 25% of agricultural employment in Kenya (Muasya and Diallo, 2001). It also accounts for nearly 20% of total food expenditures even among the poorest 20% of urban households (Muyanga

TABLE 5.1 Number of Applications by Crop (1997–2004), Kenya

Applications according to crops				
Crop type	**No.**	**Examples**	**Proportion of Residents (percent)**	**Proportion of Non-Residents (percent)**
Cereals	99	Maize, Wheat	100	0
Industrial Crop	73	Tea, Macadamia Sugarcane	100	0
Oils	31	Rapeseed, Soybean, sunflower	0	100
Pulses	20	Dry Beans	100	0
Vegetables	20	French Bean, potato	0	100
Ornamentals	247	Rose	0	100
Total	490			

et al., 2005). Dry beans are the third-most important staple food in Kenya, accounting for 9% of staple food calories and 5% of total food calories in the national diet (Ariga et al., 2010). Thus, the claim that the PVP system or the PBR system only favor crops that do not contribute to food security could not be entirely true.

On the perspective of more foreign applicants seeking protection for ornamentals, with Rose constituting about 36% of applications (1997–2004), it cannot be said that this will affect food security and sovereignty. Ornamentals are equally potential source of income into an economy and contribute the GDP of an economy through foreign exchange. Moreover, it is said "Most people don't eat flowers, but it must be appreciated that flowers are important source of food security because of the income they bring to thousands of people-most of whom are women in developing countries" (FAO, 2002), who work in these ornamental farms.

Regarding the impact of PVP/PBRs on the development of new and improved crop/plant varieties in Kenya, a recent study by UPOV (UPOV Impact Report, 2005) revealed that the Kenyan agricultural system since the inception of PVP/PBRs has seen increased introduction of foreign varieties, especially in the horticultural sector, which contribute to the diversification of the horticultural sector (e.g., the emergence of the flower industry) and support the competitiveness of Kenyan products (cut flowers, vegetables, and industrial crops) in global markets. Presently, Kenya is the largest exporter of cut flowers to Europe (Sikinyi, 2009) from Africa. Thus, based on the various reports reviewed under this study, the author can agree with Price and Lamola (1994) that without strong protection, there would be few new varieties of crop plants available for the public benefit, particularly farmers.

Another potential impact of the introduction of PBR system is the establishment of seed companies and increase in plant breeding activities. This observation was made in the case of Kenya as well. A primary rationalization supporting IPRs in general including PVP/PBRs is the incentive it provides private investment, a point empirically validated by evidence establishing the increased level of private investment. In Kenya, this could be the case based on the various reports reviewed. Plant breeding activities is also affected by the introduction of PBR. A report by UPOV (2005) indicated that breeding entities doubled from 41 pre-PVP/PBR (1990/96) to 81 in post-PVP (1997/2003). This increase also appears to be across the crop types. Consequently, the number of registered seed companies has also increased from 35 to over 50 in recent years according to Rangnekar (2006). Particularly, breeding activities have increased in recent years (1994–2003) and has resulted in a release of about 71 varieties majority of which are hybrids and by the private seed sector or breeding institutions (Ministry of Agriculture, Kenya, 2004).

According to Sikinyi (2009), staple crop production such as cassava, maize, rice, sweet potato, and wheat carried out mainly by small-scale farmers primarily to satisfy the national demand have also had an active breeding program, resulting in a number of locally bred varieties been developed and released in the food system of Kenya. This has resulted in making available to farmers high-quality seeds of improved crop varieties. Thus, the argument that, from a human rights perspective, adoption of a UPOV PVP system restricts farmers' access to seeds (Berne Declaration Report, 2014) cannot be entirely accurate. In Kenya, farmers can still save and exchange seeds but not on commercial scale. Consequently, the so-called smallholder farmers are not affected in terms of access to both their local planting materials or improved and PVP protected materials. This is corroborated by Rangnekar (2000) that farmers and growers make the choice of new, protected varieties over existing non-protected varieties, the availability of which is not affected by the PVP system. This is so because the existing nonprotected varieties remain freely available to farmers and growers after the introduction of PVP.

In terms of the economy, the introduction and implementation of a PVP/PBR system has contributed immensely. Other opportunities in implementing a PVP/PBRs system are the potential foreign exchange and job creation. The Kenyan cut flower export to the European market has increased from 129 million Euros in 1999 to 208 million Euros in 2003 (UPOV, 2005). Rikken (2011) reported it is the third largest flower exporter in the world. It is Kenya's top foreign exchange earner (Ksoll, Macchiavello and Morjaria, 2009), and thus, a source of income for the development. This enterprise also provides jobs for local people. A report by Ethical Trading Initiative (2005) indicates the cut flower industry employs over 50,000 people directly and supports several hundred thousand indirectly.

5.10 IMPACT OF PBRS IN DEVELOPING COUNTRIES: THE CASE OF SOUTH AFRICA

Although South Africa is one of the first countries to have put in place a PVP system and the first from Africa to join the UPOV system of plant protection, very little impact studies have been done. Hence, this review was limited to a Plant Breeders' Rights Policy Document, 2011, published by the Department of Agriculture, forestry and Fisheries. The author also consulted one detailed impact study by Wynand (2005) and Moephuli (2011) for the purpose of this review.

In South Africa, PBRs are recognized and protected under the Plant Breeders' Rights Act 1976 (Act no. 15). It is a member country to UPOV 1978. The Act was amended in 1996 in compliance with UPOV 1991. However, South Africa never acceded formally to the 1991 Convention, probably as the Act did not extend to all species and as Article 23A on farm-saved harvested material for re-planting, was worded in such a way that it does not comply with UPOV 1991 Article 15.2 (Wynand, 2005).

As earlier indicated, Rangnekar (2006), reported that to review the impact of PBRs' with respect to R&D, one may focus on proxy indicators like the number of applications or grants for PBRs as applied in this study.

In South Africa, 60% of PBRs holders are foreigners that are largely based in Europe and North America (Figure 5.2). This asymmetry is not unique to South Africa as a developing country. The large percentage of foreign applications may indicate the limited scope of domestic breeding activities (PBR Policy Doc. 2011). However, the database of the Registrar of PBR shows a total number of 1807 parties that have applied for PBR. Most of these are small operators, which prove that South African PBR is not dominated by a few large local or a few multinational companies (Wynand, 2005).

The implementation of plant breeders' rights system in South Africa has been a major stimulus for the plant breeding industry. Not only does it provide for financial remuneration, but also it gives local plant breeders and producers access to high quality new varieties from foreign countries (PBRs Policy Document, 2011). "If PBRs were not available locally, very few new varieties from foreign countries would be available in South Africa. This would impact negatively, among others, on the export market, as we would not be able to produce the new, sought after varieties" said a Breeder as

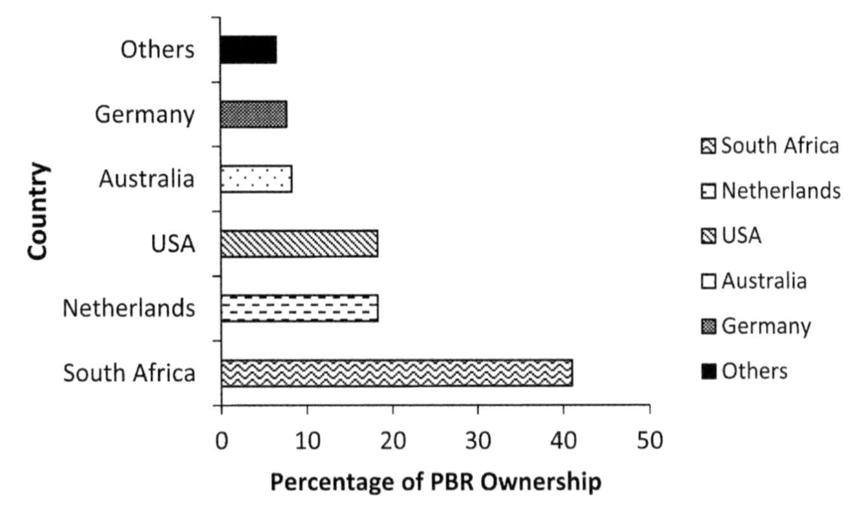

FIGURE 5.2 PBR ownership in South Africa according to countries (percent for 2009).

reported in Wynand (2005). Thus, PBR/PVP is seen among major players to be a "bait" to stimulate the development of new and improved crop plants.

From Table 5.2, it can be observed that a number of important food crops have been developed and have PBR protection. Thus, the concern regarding the tendency of research institutes shifting breeding focus from food security crops (consumable/staple) to other high-value crop plants cannot be entirely accurate. As observed in the case of Kenya, the number of ornamentals, particularly roses protected under PBRs, seems to be

TABLE 5.2 PBRs in South Africa Register in 2009

Species	Total number of PBRs in register	Percentage
Grain Maize	174	15.4
Oil Crops/Pulses	112	10.0
Fruit crops	344	30.4
Vegetable crops	45	4.0
Ornamentals (minus roses)	43	3.8
Roses	412	36.5
Total	1130	

high (452) as pertains in South Africa. It is contributing immensely to the economy of South Africa through foreign exchange earnings.

A study by Wynand (2005) demonstrated that indeed PBR plays a critical role in stimulating crop development by both public and private research institutions and seed companies. The study compared PBRs granted over 4 years from 2001–2004 against a total of over 15 years since 1990. It was revealed that for wheat, 26 (45%) of the total PBRs granted since 1990, were granted in the past 4 years; for potatoes 25 (41%) in the last 4 years; for dry beans 20 (43%) in the last 4 years; and for soybeans 44% in the last 4 years. This is the period when the PBRs legal framework in South Africa saw a review toward UPOV-91 system compliance.

Figure 5.3 shows that provided there is incentive through legal protection, various players in the seed and research industry are willing to invest in crop development/plant breeding activities. For instance, private seed companies and breeders own almost 50% of the 704 varieties under the South African PBR system. Wynand (2005) again reported that in terms of both public–private breeding activities, the private South African breeders and companies own 431 varieties, while 237 varieties were for public institutions, showing the level of investment by the private sector in plant breeding. This also demonstrates the benefit of PBR for successful investment in breeding by local private breeders.

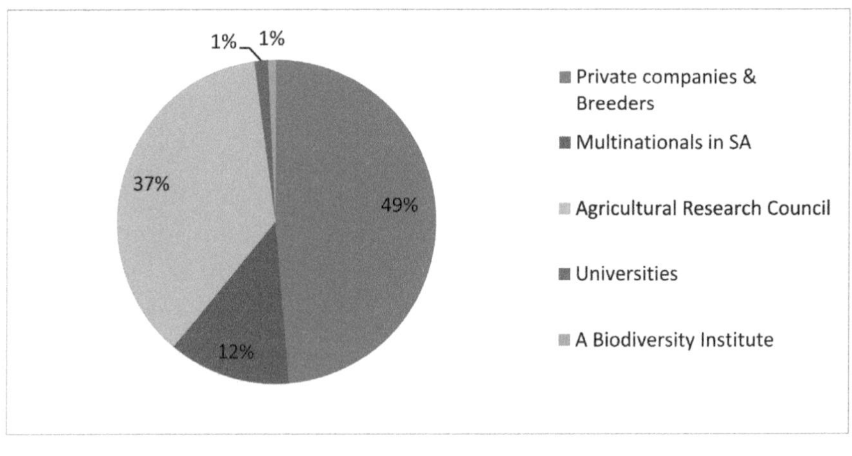

FIGURE 5.3 Category of PBRs ownership in South Africa.

5.11 MANAGEMENT OF AGRICULTURAL RESEARCH AND DEVELOPMENT (R&D) OUTPUTS: PUBLICLY FUNDED RESEARCH INSTITUTIONS

Agricultural R&D, most of which is supposed to be publicly funded, play a vital role in developing national innovative capacity for economic growth. With the increasing changing trend regarding the contribution of the private sector in research and development, the application of modern biotechnology among others, the protection of IPR, particularly in agriculture, has become imperative. The complex question faced by most publicly funded R&D institutions is how to harness the IP owned by others and how to protect their own generated technologies. Majority of the public institutions in developing countries lack the human resource and financial capacity to manage their own IP.

One major complex task to overcome is the notion that public research institutes are to generate knowledge, technologies, and products for "public good." Unfortunately, however, most people have misconstrue this notion to be public research institutes must make these knowledge and technologies available to the general public at no cost. Therefore, the notion of protecting one's innovations, especially plant varieties and animal technologies seems at odds with the mission of public research (Maredia et al., 1999). Thus, one of the questions facing public institute managers and researchers is whether they should protect their intellectual property? In as much as there could not be a straight forward answer, it is generally agreed that some level of protection encourages more innovations and creativity. The most important mechanisms for legally protecting agricultural innovations as enshrined in the TRIPs Agreement are plant variety rights/plant breeders' rights and patents. Other forms could be use of trademarks, trade secrets, and copyrights. MTAs are also increasingly been used by public research institutes both at national and international levels (Cohen et al., 2000).

After securing propriety rights over such agricultural R&D outputs, the public institute exploits the IP asset by licensing to others to generate revenue. The technology can be given to a private firm to exploit it at an agreed rate of royalty payments. However, the viability of this option depends on how successful the private firm is able to market the technology. The second option could be using as a bargaining chip to negotiate other technologies of interest from private sector. The case of EMBRAMPA in Brazil is a classic example (Sampaio, 1999). Other options of setting up spin-off companies exist; however, this usually requires huge funding, which often is unavailable for the public institution. Thus, in this regard, public–private partnership could be negotiated for such ventures.

5.12 CONCLUSION

In conclusion, there is empirical evidence to show the role PVP or PBRs play in ensuring food security, improving malnutrition, and reducing poverty across the world. Looking into the future, with the growing human population and the myriad challenges posed by changing climatic conditions, the protection of agricultural R&D outputs has become necessary. This will encourage private sector investment, spur innovation, and enhance creativity to address the current and future challenges. PBRs as one of the IPR have proven to be the most suitable mechanism so far in protecting agricultural research. That notwithstanding, the adoption and implementation of a PBR system must take into account the national agricultural topography. The systems of agriculture in the developed countries are not the same for countries in the developing and least developed world. Hence, PBRs for individual countries must reflect the national context and vision for agricultural R&D.

KEYWORDS

- **agriculture**
- **food**
- **ITPGRFA**
- **patent**
- **PVP**
- **UPOV**

REFERENCES

Agreement on trade-related aspects of Intellectual Property Rights (1994). Marrakech, 33 *International Legal Materials*.

Akroyd, S., & Smith, L., (2007). Review of public spending to agriculture. *A Joint DFID/World Bank Study*, Oxford Policy Management, Final Draft, January, 2007, 25–61.

Alston, J., & Venner, R., (2002). The effects of the, US plant variety protection act on wheat genetic improvement. *Research Policy, 31*, 527–542.

Ariga, J., Jayne, T. S., & Stephen, N., (2010). Variation in staple food prices: Causes, consequence, and policy options, *Proceedings of the Seminar on "Variation in Staple Food Prices: Causes, Consequence, and Policy Options," Maputo, Mozambique*, COMESA-MSU-IFPRI, African Agricultural Marketing Project (AAMP), 25-26 January, 2010.

Barton, J. H., (1998). The impact of contemporary patent law on plant biotechnology research. In: *Global Genetic Resources: Access and Property Rights*. CSSA Special Publication, Madison: Crop Science Society of America, 85–97.

Barton, J. H., (2006). '*Knowledge,' Expert Paper Series Six: Knowledge*, Secretariat of the International Task Force on Global Public Goods, Stockholm, 1–20.

Bently, L., & Sherman, B., (2009). *Intellectual Property Law*, 3rd ed. Oxford University Press.

Berne Declaration Owning Seeds (2014). Accessing food: A human rights impact assessment of UPOV 1991, based on case studies in Kenya, Peru and the Philippines, 1–56.

Blakeney, M., Cohen, J. I., & Crespi, S., (1999). Intellectual property rights and agricultural biotechnology. In: *Managing Agricultural Biotechnology Addressing Research Program Needs and Policy Implications,* Cohen, J. I., (ed.), Wallingford, UK: CABI Publishing, 207–227.

Bortey, H. M., & Mpanju, F., (2016). Adoption of plant breeders' rights system: Perceived implication for food, seed security and sovereignty in Ghana. *Journal of Intellectual Property Rights, 21*, 96–104.

Butler, L., (1996). Plant breeders' rights in the, U. S.: Update of a 1983 study. In: *Proceedings of a Seminar on the Impact of Plant Breeders' Rights in Developing Countries*, Van Wijk, J., & Jaffe, W., (ed.), University of Amsterdam, Amsterdam, pp. 17–33.

Canadian Food Inspection Agency, (2001). 10-Year review of Canada's plant breeders' rights act, Ottawa, Canada.

Cohen, J. I., Falconi, C., & Komen, J., (2000). Perspectives from international agricultural research centers, In: *Intellectual Property Rights in Agriculture: The World Bank's Role in Assisting Borrower and Member Countries*, Uma Lele, William Lesser, & Gesa Horstkotte-Wesseler, (ed.), Washington, DC: The World Bank, pp. 107.

Conway, G., (1999). *The Doubly Green Revolution – Food for All in the Twenty-First Century*. First Edition, Cornell University Press.

Cornish, W., Llewelyn, D., & Aplin, T., (2013). *Intellectual Property: Patents, Copyright,* Trademarks *and Allied Rights*, 8th ed. Sweet & Maxwell, Research Collection School of Law, pp. 1–945.

Cullet, P., (2001). Plant variety protection in Africa: Towards compliance with the TRIPS agreement. *Journal of Africa Law*, *45*(1), 97–122.

Cullet, P., (2003). Plant variety protection: Patents, plant breeders' rights & sui generis systems. *IELRC Briefing Paper, 4*. Available at: http://www.ielrc.org/content/f0304.htm.

Diez, M. C. F., (2002). The impact of plant varieties rights on research: The case of Spain. *Food Policy, 27*(2), 171–183.

Donnenwirth, J., Grace, J., & Smith, S., (2004). Intellectual property rights, patents, plant variety protection and contracts: A perspective from the private sector. *IP Strategy Today No. 9*, 19–34.

Dutfield, G., (2011). Food, biological diversity and intellectual property: The role of the international union for the protection of new varieties of plants *(UPOV)* QUNO, *Intellectual Property Paper No. 9,* QUNO, Geneva.

Ethical Trading Initiative (ETI), (2005). *Addressing Labour Practices on Kenyan Flower Farms*. London. Retrieved March 2, (2013). from:http://www.ethicaltrade.org/resources/key-eti-resources/addressing-labourpractices-on-ke.

FAO, (2014). IFAD and WFP. The State of Food Insecurity in the World 2014. Strengthening the enabling environment for food security and nutrition. Rome, FAO. Available at:http://www.fao.org/3/a-i4030e.pdf.

FAO-Global agriculture towards 2050 high expert forum report, (2009). Available at: http://www.fao.org/fileadmin/templates/wsfs/docs/Issues_papers/HLEF2050_Global_Agriculture.pdf.

FAO-Global Meeting to assess progress on World food Summit Goals', (2001). Available atwww.fao.org/news).

FAO-Incorporating food security concerns in a revised agreement on agriculture (2001), FAO round table on food security in the context of the WTO negotiations on agriculture. *Discussion Paper No. 2*. Available at: http://www.fao.org/docrep/005/Y3733E/y3733e03.htm.

FAO-The State of Food and Agriculture: Lessons from the past 50 years, FAO, Rome, (2000). Available at: http://www.fao.org/docrep/x4400e/x4400e00.htm.

FAO-The state of food insecurity in the world, (2002), FAO, Rome, (2002). Available at:http://www.fao.org/docrep/005/y7352e/y7352e00.htm.

GRAIN, (1999a), Plant variety protection to feed Africa? 25 Seedling-December. Available at:https://www.grain.org/article/entries/230-plant-variety-protection-to-feed-africa.

IPGRI, (1997). Options for a *sui generis system*. No 6, Issues in Genetic Resources, Rome.

Jordens, R., & Button, P., (2011). Effective system of plant variety protection in responding to challenges of a changing world: UPOV Perspective. *Journal of International Property Rights, 16*, 74–83.

Kameri-Mbote, P., Odote, C., Musembi, C., & Kamande, M., (2013). *Ours by Right: Law, Politics and Realities of Community Property in Kenya,*"Strathmore University Press.

Kanwar, S., & Evenson, R., (2001). *Does Intellectual Property Protection Spur Technological Change?* Oxford Economic Papers, Oxford University Press, vol. *55*(2), pp. 235–264.

Ksoll, C., Macchiavello, R., & Morjaria, A., (2009). *Guns and Roses: The Impact of the Kenyan Post-Election Violence on Flower Exporting Firms*. CSAE WPS/2009–06.

Langford, J., (1997). Intellectual property rights: Technology transfer and resource implications. *American Journal of Agricultural Economics, 79*(5), 1576–1583.

Leenders, H., (1976). The European view on the EEC certification rules. *Agricultural Merchant, Seeds Special, 31*, *56*(7), pp. 33–34.

Leskien, D., & Flitner, M., (1997). Intellectual property rights and plant genetic resources: Options for a sui generis system (Rome). International Plant Genetic Resources Institute, *IPGRI, Issues in Genetic Resources, 6*, pp. 77.

Lesser, W. H., (1990). Sector issues II: Seeds and plants'. In: *Strengthening Protection of Intellectual Property in Developing Countries*, Siebeck, W. E., Evenson, R. E., Lesser, W., & Primo, B., Carlos, A., (ed.), Washington, DC, USA., The World Bank, 59–68.

Lesser, W., (1997). Assessing the Implications of intellectual property rights on plant and animal agriculture. *American Journal of Agricultural Economics, 79*(5), 1584–1591.

Louwaars, N. P., Trip, R., Eaton, D., Henson-Appollonio, V., Hu, R., Mendoza, M., Muhuuku, F., Pal, S., & Wenkundah, J., (2005). *Impact of Strengthening IPR Regimes on the Plant Breeding Industry in Developing Countries:* A synthesis of five case studies, World Bank Propress, Wageningen, Netherlands.

Louwaars, N., Dons, H., Van Overwalle, G., Raven, H., Arundel, A., Eaton, D., & Nelis, A., (2009). *Breeding Business:* The future of plant breeding in the light of developments in patent rights and plant breeder's rights, CGN Report 2009-14, Wageningen, Centre for Genetic Resources.

Maredia, K. M., (2001). *Application of Intellectual Property Rights in Developing Countries:* Implications for public policy and agricultural research institutes. World Intellectual Property Organization, WIPO. Final Draft submitted to WIPO, pp. 1–69.

Maredia, M., Howard, J., Boughton, D., Naseen, A., Wanzala, M., & Kajisa, K., (1999). *Increasing Seed System Efficiency in Africa: Concepts, Strategies and Issues*, Michigan State University, International Development Working Paper No. 77, Department of Agricultural Economics- MSU East Lansing Michigan. Available online at: http://www.aec.msu.edu/fs2/papers/idwp77.pdf.

Merges, R. P., (2011). *Justifying Intellectual Property*, Harvard University Press, Cambridge, Massachusetts London, England.

Merges, R., Menell, P., & Lemley, M., (2003). *Intellectual Property in the New Technological Age*, New York: Aspen Publishers.

Moephuli, S. R., (2011). *Agricultural Research: Enhancing Productivity through Intellectual Property*, WIPO Seminar. http://www.wipo.int/edocs/pubdocs/en/global_challenges/1027/wipo_pub_1027.pdf, pp. 36–59.

Morris, M., & Ekasingh, B., (2002). Plant breeding research in developing countries: What roles for the public and private sectors? In: *Agricultural Research Policy in an Era of Privatization*, Derek, B., & Ruben, E., (ed.), Oxon and New York, CAB International, 199–225.

Morris, M., Edmedes, G., & Pehu, E., (2006). The global need for plant breeding capacity: What roles for the public and private sector. *Hort. Science, 41*(1), 30–39.

Muasya, W. N. P., & Diallo, A. O., (2001). Development of early and extra early drought and low nitrogen -tolerant varieties using exotic and local germplasm for the dry mid -altitude ecology. In: Friesen, D. K., & Palmer, A. F. E., (eds.). Integrated approaches to higher maize productivity in the new millennium. *Proceedings of the Seventh Eastern and Southern Africa Regional Maize Conference*, Nairobi, Kenya, CIMMYT and KARI, 253–259.

Muchiri, W. C., (2014). Plant breeders' rights in Kenya: Examining the effect of DUS test in food crops and ornamentals. *LLM Dissertation*, University of Nairobi, Kenya, pp. 1–59.

Murphy, P. W., (1980). Evidence submitted by Mr., Murphy, P. W., School of Law, Controller of the United Kingdom plant variety rights office, to the subcommittee of department of investigations, oversight and research of the house of representatives agriculture committee of the united states of America. *UPOV Newsletter, 23*, 3–8.

Muyanga, M., Jayne, T., Kodhek, G., & Ariga, J., (2005). St*aple Food Consumption Patterns in Urban Kenya: Trends and Policy Implications.* Working Paper 19, Tegemeo Institute, Nairobi. http://www.tegemeo.org/documents/work/tegemeo_workingpaper_19.pdf.

Narasimhan, S. M., Hans, M. H., & Manuel, R. M., (2011). Food security and intellectual property rights finding the linkages. In: *Intellectual Property and Human Development*. Tzen Wong, & Dutfield, G., (eds.). Cambridge University Press, pp. 103–131.

Oxfam. Rigged rules and double standards, (2002): trade, globalisation and the fight against poverty' Oxford, Oxfam. Available at:http://en.oxfam.ru/upload/iblock/1ca/1ca3150c5e7d8c64ba45287978f0d4df.pdf.

Pardey, P. G., Alston, J. M., & Smith, V. H., (1997). Financing science for global food security, *IFPRI Annual Report Essays*, Washington, DC. Available at: http://www.cgiar.org/ifpri/pubs/books/ar1997–1.htm.

Penna, A. L. d. R., (1994). An analysis of the impact of plant breeders' rights legislation on the introduction of new varieties in UK horticulture. PhD Dissertation, Wye College, University of London.

Plant Breeders' Rights Policy of South Africa, (2011). Department of Agriculture, Forestry and Fisheries. Available at: http://www.daff.gov.za/docs/Policy/PlantBreederPol_2011.pdf.

Price, S. C., & Lamola, L. M., (1994). Decision Points for transferring plant intellectual property to the private sector. In: *Conservation of Plant Genes II: Utilization of Ancient and Modern DNA, Monograph in Systematic Botany 48*. Adams, R. P., Miller, J. S., Golenberg, E. M., & Adams, J. E., (ed.), St. Louis: Scientific Publications.

Rangnekar, D., (2000). *Intellectual Property Rights and Agriculture*: An analysis of the economic impact of plant breeders' rights. London, Action-aid. https://www.actionaid.org.uk/sites/default/files/doc_lib/ipr_agriculture.pdf, pp. 1–58.

Rangnekar, D., (2002). *Access to Genetic Resources*, gene-based inventions and agriculture "Commission on Intellectual Property Rights Background Paper, 3a, Commission on Intellectual Property Rights," London, pp. 39. Source: http://www.iprcommission.org.

Rangnekar, D., (2006). Assessing the economic implications of different models for implementing the requirement to protect plant varieties: A case of Kenya. A report prepared under the European Commission's 6th Framework Programme for Research (Contract No. 503613). *Impacts of the IPR Rules and Sustainable Development (IPDEV).* https://www.ecologic.eu/sites/files/download/projekte/1800-1849/1802/wp6_india_case_study.pdf, pp. 1–55.

Repetto, S. R., & Cavalcanti, M., (2009). Article 27.3(b): *Related International Agreements* (Part I), module 5 IV, agreement on trade-related aspects of intellectual property rights (TRIPS) (FAO Corporate Document Repository).

Rikken, M., (2011). The Global Competitiveness of the Kenyan Flower Industry. Prepared for the Fifth Video Conference of the Flower Industry in Eastern Africa.

Rosegrant, M. W., Paisner, M. S., Meijer, S., &Witcover, J., (2001). Global food projections to 2020: *Emerging Trends and Alternative Futures*. Washington, DC. IFPRI.

Washington, D.C., http://www.amazon.com/Global-Food-Projections-2020-Alternative/dp/0896296407, pp. 1–17.

Sampaio, M. J. A., (1999). Perspectives from national systems and universities: Brazil. In: *Intellectual Property Rights in Agriculture: The World Bank's Role in Assisting Borrower and Member Countries*. Uma Lele, William Lesser, & Gesa Horstkotte-Wesseler, ed., Washington, D. C., The World Bank, pp. 44–66.

Schroeder, C., Onyango'Oloo, T., Ranabhat, N. B., Jick, N. A., Parzies, H. K., & Gemenet, D. C., (2013). Potentials of hybrid maize varieties for small-holder farmers in Kenya: A review based on SWOT analysis. *African Journal of Food, Agriculture, Nutrition and Development*, *13*(2), 7562–7586.

Serageldin, Ismail, & Persley, G. J., (2000). *Promethean Science*. Agricultural biotechnology, the environment and the poor, Consulting Group for International Agricultural Research, Washington DC, pp. 48.

Sikinyi, E., (2009). Plant variety protection (Plant Breeder's Rights) in Kenya. In: *Intellectual Property Rights in Kenya*, MoniWekesa, & Ben Sihanya, (ed.), 73–107.

Singh, H., (2007). Plant variety protection and food security: Lessons for developing countries. *Journal of Intellectual Property Rights*, *12*, 391–399.

Srinivasan, C. S., (2004). Plant variety protection, innovation and transferability: Some empirical evidence. *Review of Agricultural Economics, 26*(4), 445.

Tabor, S. R., (1998). Trends in agricultural research funding. In: *Financing Agricultural Research: A Sourcebook*, Tabor, S. R., Janssen, W., & Bruneau, H., (ed.), The Hague, ISNAR, pp. 301.

Tripp, R. N., (2007). Louwaars and Derek Eaton. Plant variety protection in developing countries. A report from the field. *Food Policy, 32*, 354–371.

UPOV, (2005). *International Union for the Protection of New Varieties of Plants*, UPOV report on the impact of plant variety protection, Geneva. http://www.upov.int/edocs/pubdocs/en/upov_pub_353.pdf, pp. 84–86.

Watal, J., (2001). *Intellectual Property Rights in the WTO and Developing Countries* (London: Kluwer Law International, 136–149.

Wekesa, E., Mwangi, W., Verkuijl, H., Danda, K., & De Groote, H., (2003). *Adoption of Maize Production Technologies in the Coastal Lowlands of Kenya*. CIMMYT, Mexico, D. F., pp. 1–22.

Wekesa, M., Muraguri, L., & Boadi, R., (2009). IPRs, *Agriculture and Food Security in Intellectual Property Rights in Kenya*. Wekesa, M., & Sihanya, B., (eds.), pp. 39–67.

Wekesa, Moni, (2005). Internationalization of intellectual property rights through information communication technologies, In: *Re-invigorating the University Mandate in a Globalizing Environment: Challenges, Obstacles and Way Forward*. DAAD Regional Office, Nairobi, 121–130.

WIPO, (2004). *Intellectual Property Handbook Policy, Law and Use*. WIPO Publication No. 489 (E).

Wynand, J., & Van Der Walt, (2005). A study on status of plant variety protection in the SADC region country reports for South Africa, Angola, Malawi, Mozambique, Zambia and Zimbabwe. *FANRPAN Economic Policy Research*, 1–57.

CHAPTER 6

FARMERS' RIGHTS AND PRIVILEGES: IMPLICATION FOR THE FARMING COMMUNITY

ARPITA DAS,[1] BHOLANATH SAHA,[2] P. K. BHATTACHARYA,[3] and ANITA ROY[3]

[1] *Bidhan Chandra Krishi Viswavidyalaya, Mohanpur, Nadia, West Bengal, India, E-mail: arpitacoh@gmail.com*

[2] *Dr. Kalam Agricultural College, Bihar Agricultural University, Kishanganj, Bihar, India*

[3] *Bidhan Chandra Krishi Viswavidyalaya, Mohanpur, Nadia, West Bengal, India*

CONTENTS

ABSTRACT

India has established an intellectual property rights regime that covers plant varieties and enacted "The Protection of Plant Varieties and Farmers' Rights Act 2001" that seeks to protect the rights of farmers and breeders on plant varieties. Within this framework, India has chosen to put in place a *sui generis* system, one that has the declared objective of protecting farmers' rights. This act recognizes the role of individual farmers and farming community in the improvement and conservation of varieties and traditional landraces. The Protection of Plant Varieties and Farmers' Rights (PPV & FR) Act is the legislative expression of this commitment of the Indian state. This chapter reviews the PPV & FR Act in the perspective of farmers' right and privilege status of farmers' variety registration and implication of this act in the farming sector in India.

6.1 INTRODUCTION

The importance of farmers cannot be undermined in any country in general, and in developing countries in particular, as they need special protection system due to more dependence on farming for economic growth and development. From time immortal, farmers around the globe have been involved in domestication, selection, and conservation of germplasms of different crops. Continuous breeding effort of farmer through various approaches have created a rich wealth of varieties in many crops. This wealth consists of landraces, wild species, obsolete varieties, and old and new varieties, which are the foundation of variability that scientists can use for improvement of existing varieties or development of new varieties. India is the center of origin of many crop species like rice, little and kodo millet, red gram, mung bean, bengal gram, pepper, cardamom, many vegetables, and fruit species. Moreover, many crops like wheat, sorghum, maize, pearl millet, ragi, ground nut, sugarcane, cotton, tea, rubber, soybean, and sunflower, and many important horticultural crops have been introduced in this country. During the long process of domestication, selection, conservation, and cultivation, farmers have acquired knowledge regarding the concept of plant breeding and created and conserved the genetic diversity. The genetic diversity present in traditional varieties is most useful and economically valuable part of global biodiversity. Traditional varieties (landraces) are directly used by subsistence farmers as a key component of their cropping systems. Such varieties account for about 60% of agricultural land use and provide approximately 15% to 20% of the world's food (Wood and Lenne, 1997). In addition, landraces are the basic raw materials used by plant breeders for the development of modern varieties, which provide the remainder of the world's crop production. Rapid expansion of modern agricultural technology has caused disruption of natural systems, leading to genetic erosion. The conservation and sustainable utilization and access to biological diversity were considered as national sovereignty by Convention on Biological Diversity (CBD). Consequently, many issues regarding the rights of the conservers, users, breeders, farmers, and intellectual property have emerged (Brahmi and Dhillon, 2004).

While the contribution of scientists in the advancement of agriculture is measurable and quantifiable in the form of desired outcomes, the contribution of large number of individual farmers and farming communities who are engaged in agriculture for generations after generation in conserving the biodiversity which many times has even contributed in the development of many improved varieties goes unnoticed (Kochar, 2008). A new legislation called *The Protection of Plant Varieties and Farmers' Rights (PPV & FR) Act 2001* enacted by the Government of India, therefore, seeks to protect the rights of farmers and breeders on plant varieties. This Act recognizes the farmers and farming community that plays an important role in the improvement and conservation of varieties. As farming is a family activity, individual and community rights of farmers mean equitable rights for men and women engaged in farming (Bala Ravi, 2004).

6.2 FARMERS' RIGHTS: RECOGNITION

The first use of Farmers' Rights as a political concept dates back to the early 1980s, when Rural Advancement Foundation International coined the term Farmers' Rights (FR) to highlight the valuable but unrewarded contributions of farmers to conserve plant genetic resources (Ramanna, 2006). In recognition of this critical factor, an independent commission on plant genetic resources, International Undertaking on Plant Genetic Resources for Food and Agriculture (IUPGRFA), constituted in 1983 by the Food and Agriculture Organization, Rome, under the chairmanship of Prof. M.S. Swaminathan, first formally introduced the concept of FR (Seema, 2012). Consequently, the group agreed, the recognition of not only the individual farmers or communities of farmers but the rights of entire peoples who, continuously engaged in conservation, maintenance, and improvement of plant genetic resources silently. FR defined by the IUPGR were recently made into a legally binding international treaty. This treaty is called International Treaty on Plant Genetic Resources for Food and Agriculture (ITPGRFA). India is a party to this treaty and hence committed to protect farmers' rights on seeds and genetic resources. The IUPGR affirmed the recognition of farmers for improvement and conservation of genetic diversity, acknowledgment of farmers' recognition through monetary incentive, and their contribution in the field of crop improvements by breeding and other scientific methods (Kochupillai, 2011). In respond to international development, India also initiated the effort to establish a FR regime to recognize and protect the right's arising from the past, present, and future contributions of farmers in conserving, improving the plant genetic resources. In India, agriculture provides employment and key means of livelihood to more than 65% of population. Among them a vast majority of the farming population constitute small and marginal farmers (Sahay and Srivastava, 2001).

6.3 HISTORY AND GENESIS OF INDIAN LEGISLATION FOR FARMERS' RIGHT

The legally valid ownership rights assigned to the creator of intellectual work are called intellectual property rights (IPR). Intellectual property related to farmers' and breeders' creation is a big issue. Ownership right is established on the basis of evidence that the person has contributed to the development and conservation of the variety (Chandrasekharan and Vasudev, 2002).

Before the existence of any international convention, it was difficult to obtain protection in many countries due to the diversities in the national laws. Globalization necessitated harmonization of industrial law. Intellectual property protection provided by countries varied, and this disparity was an occasion for developing countries to pirate the technological development. The developed countries under the leadership of United States wanted to have a strong system for the protection of new technologies. Hence, the issue of intellectual property protection and also protection of new plant varieties was addressed in Uruguay Round of General Agreement on Tariff and Trade. The minimum requirement for protection of intellectual property in plants varieties is defined by TRIPS Agreements. Plant varieties are only referred to once in the TRIPs Agreement, according to which all WTO member states have to adopt legislation for the effective implementation of plant variety protection, including enforcement (Gopalkrishnan, 2004). In order to be in conformity with the TRIPS Agreement, there must be in place an effective *sui generis* system for plant varieties. According to the TRIPS Agreement, for a genetically modified plant variety, a person would be entitled to a patent and in other cases he/she should be entitled to protection under a *sui generis* system. There is no definition of "effective sui generis system." When a TRIPS Council issued a document, consisting compilation of the responses to a questionnaire within the scope of the review of Article 27.3(b), the term effective was not applied. The development of plant variety protection seemed to have provided the appropriate solution to the question of industrial property protection for plants, and there was little debate on the suitability of the system.

In India, agricultural research including the development of new plant varieties has largely been the concern of the government and public-sector institutions (Brahmi et al., 2004). Earlier, India did not have any legislation to protect the plant varieties and, in fact, no immediate need was felt. However, after India became signatory to the TRIPS Agreement in 1994, such a legislation was necessitated. Article 27.3 (b) of this agreement requires the member countries to provide for protection of plant varieties either by a patent or by an effective sui generis system or by any combination thereof. The existing Indian Patent Act 1970 excluded agriculture and horticultural methods of production from patentability. The *sui generis* system for the protection of plant varieties was developed integrating the rights of breeders, farmers, and village communities, and taking care of the concerns for equitable sharing of benefits (Roy, 2013). It offers flexibility with regard to protected genera/species, level and period of protection, when compared to other similar legislations existing or being formulated in different countries. The Act covers all categories of plants, except microorganisms. The genera and species of the crop for protection shall be notified through a gazette, after the appropriate rules and by-laws are framed for the enforcement of the Act. Convention of Biological Diversity (CBD, 1993) regulates the conservation, sustainable utilization, and access to the biological diversity in India, whereas the PPV & FR Act (2001) regulates the Researchers' Rights, Breeders' Rights, and Community Interests on cultivated crops.

6.4 PROTECTION OF PLANT VARIETIES AND FARMERS' RIGHT ACT 2001

In 2001, the Indian Parliament passed an act named "Protection of Plant Varieties and Farmers' Rights Act (PPVFR Act, 2001)" and its rules were drafted in 2003. A *sui generis* system (system of own) is an attempt by the Indian Government to recognize and protect the rights of both commercial plant breeders and farmers in respect of their contribution made in conserving, improving, and making availability of plant genetic resources for the development of new plant varieties. This act formulated with the following objectives:

- To provide an effective system for the protection of plant varieties and to ensure plant breeders' rights to stimulate investment for research and development and continuous development of new varieties.
- To protect farmers' varieties and recognize the farmers in respect of their contribution made at conserving, improving, and making availability of plant genetic resources for the development of new plant varieties.
- To facilitate the growth of the seed industry to ensure production and availability of high-quality seed/planting material.

For proper implication of the PPV & FR Act, India established the Protection of Plant Varieties and Farmers' Rights Authority on November

11, 2005 (PPV & FR Authority, 2005) vide Gazette Notification No. S.O. 1589(E). The Authority is a body corporate consists of a Chairperson and 15 members with head quarter at New Delhi and branch offices at Guwahati and Ranchi.

It is a well-known fact that farmers have been associated in conserving the agro-biodiversity which is linked with varied agro-climatic and social conditions. However, it is an irony that the farmer's role in this aspect as a conserver or improver does not always recognized properly due to the economic compulsions of the farming community. Keeping in view the past, present and future contribution of the farming communities in agro-biodiversity conservation, the PPVFR Act provides farmers' rights including benefit sharing, compulsory licensing, compensation, and other provisions which effectively create a framework for recognition and rewarding.

6.5 WHO IS THE FARMER?

A farmer is widely understood to be the person husbanding crops and animals; the definition given in this Act is very unique. According to Section 2(k) of the act "Farmer" means any person who cultivates crops by cultivating the land himself or directly supervising for cultivation of land through any other person as well as conservation and preservation of any wild species or traditional varieties individually or jointly along with any person and value addition of such wild species or traditional varieties through selection and identification of their useful properties as well. In other words, the Act recognizes the farmer as a cultivator, conserver, and breeder. This definition embraces all farmers, landed or landless, male and female.

6.6 WHAT IS A FARMERS' VARIETY?

A variety that has been traditionally cultivated and evolved by the farmers in their fields or is a wild relative or landrace of a variety about which the farmers possess common knowledge is known as farmers' variety (FV) (Brahmi, 2004). All plant varieties, except modern improved varieties, have been selected, improved, and conserved by farmers. Depending on the level of improvement and use in cultivation, they are called traditional varieties, folk varieties, and landraces (Nagarajan, 2010). All of them are essentially FVs. Apart from the biological entities constituting these varieties, the rich knowledge that farmers have about each of them is equally important. This knowledge is the decisive factor for assessing the importance of these varieties to agriculture, in the present and future. This knowledge has been generated and conveyed over long years of intelligent observations and improvements effected on these varieties by farmers across different growing conditions. During the process of selection, cultivation, and conservation of these varieties across hundreds of years, farmers have acquired knowledge on the unique properties of each variety and how these properties could be used in agriculture. The uniqueness and usefulness of this knowledge is far more valuable than the innovation and utility assigned on a new variety for granting a patent or plant breeders rights. The knowledge of farmers goes beyond all cultivated varieties to the wild relatives of crop plants, which naturally grow around. The PPV & FR Act recognizes these wild species as FVs, if the farmers possess useful knowledge on them (Srinivas, 2003).

6.7 FARMERS' RIGHTS AND PRIVILEGE

The Act seeks to institutionalize the public interest on conservation by provisioning reward and recognition to individual farmers and community conservers. Overall, this Act provides nine important privileges and rights to the farmers. The prominent descriptions of the Act in this regard are as follows:

6.7.1 FARMERS' RIGHT ON SEED

The farmers' right on seeds is a traditional right enjoyed by farmers all along the history of agriculture. This right includes to save their own seed produced by him and to use for sowing, re-sowing, exchanging, sharing, and selling to other farmers. Therefore, this act includes all the varieties including protected varieties. However, farmers are restricted to sell seed using brand name of a protected variety according to Section 39 (1), (i)–(iv) of the act. To further safeguard this right on the seeds of registered varieties, the act prohibits the use of technologies like the terminator gene technology or any genetic use restriction technology (GURTs), which restricts the repetitive use of saved seeds [Annexure 2, Section 18 (1) (c)].

6.7.2 FARMERS' RIGHTS TO REGISTER VARIETIES

The PPV & FR Act allows the registration of traditional varieties or FVs [Section 14]. A farmer who bred or developed a new variety shall be entitled for registration and other protections as like new varieties developed by professional breeders. Therefore, this act recognizes the role of farmer for arresting genetic erosion as well as conservation of genetic resources and further improvement just like a breeder. In case of varieties where there is information gap regarding the role of an individual to evolve that particular variety, the variety can be registered in the name of farming community of the region instead of individual farmers/community.

6.7.3 FARMERS' RIGHT FOR REWARD AND RECOGNITION

The farmer(s) who is engaged in conservation of the genetic resources of cultivated and wild relatives of economic plants and their improvement through selection and preservation shall be entitled in the prescribed manner for recognition and reward from the national gene fund, provided that material so selected and preserved has been used as donors of genes in new varieties registerable under Sections 39(1) (iii).

6.7.4 FARMERS' RIGHT FOR BENEFIT SHARING

Recognizing the role of farmers' for improving or developing new plant varieties is done through benefit sharing. An entrepreneur should not ignore the contribution of a farmer to evolve the variety while commercialization. The breeder of any registered variety who developed FV as a parent should ensure benefit sharing with the farmer or community [Section 26 (1)] through depositing the amount in the Gene Fund for the purpose of payment to the claimant. The Act also grants right of ownership of "Original Variety" of an essentially derived variety (EDV): a mandatory need to secure consent of farmer(s) when an FV is used to develop an EDV. Normally, all applicants who seek to register new varieties are required to declare the source of the varieties used as parents for breeding new varieties [Section 18 (e), (h)]. This Act gives opportunity to the farmer or community to claim benefit shared by the professional breeder who has used the FV as a parent to breed his variety [Section 26 (2)]. Claims for benefit sharing have to be made in Form PV-7 prescribed by the Authority within 6 months from the date of advertisement inviting such claims by the Authority. On examination of such claims by the Authority applicant eligible for benefit sharing and the quantum of benefit to be shared are decided. The PBR holder of the variety is required to remit the awarded benefit share in the National Gene Fund [Section 26 (6)]. The benefit share may be disbursed from the National Gene Fund to the eligible individual, community, or institution [Section 45 (2) (a)]. Therefore, farmers must be well aware about the information regarding registration and notification of a new variety and about parentage of this variety for claiming benefit sharing. Information regarding varietal notification followed by awareness generation can be promulgated through linking farmer's cooperative, village line departments, and gram panchayats with the NGOs and agricultural Institutes.

6.7.5 FARMERS' RIGHT TO COMPENSATION FOR LOSS SUFFERED FROM REGISTERED VARIETY

The basic purpose of registration of any variety under the PPV& FR Act is ensuring the commercial right that encourages the demand of the variety. The expected performance of a variety is to be disclosed to the farmers at the time of sale of seed/propagating material. The Act entitles the farmers' right to enjoy a claim for compensation for under performance of a right protected variety from its promised level under defined production conditions [Section 39 (2)]. Such claims for compensation are made to the PPV & FR Authority in Form PV-25. For compensation claiming, farmer or farming community has to ensure that the cultivation of that variety is carried out following recommended package of practices. The Authority on confirmation of the compensation claim decides the amount of compensation to be paid by the PBR-holder.

6.7.6 FARMERS' RIGHT FOR PROTECTION AGAINST INFRINGEMENT

Under legal jurisprudence, violation of a law committed out of ignorance is not held as an admissible innocence (Kanwar and Evenson, 2003). A safeguard to farmers against innocent infringement is provided in the Act [Section 42]. Considering poor legal literacy prevailing among farmers and to discourage legal harassment through infringement proceeding, a safeguard to farmers against innocent infringement is provided in the Act. A right established under this Act shall not be deemed to be infringed by a farmer who proves that at the time of such infringement that he was not aware of the existence of the right so infringed.

6.7.7 FARMERS' RIGHT FOR RECEIVING FREE SERVICES

Considering the financial crunch of farmer, the PPV & FR Act provides opportunity to the farmer for free service, so that their poor economic status shall not be an obstacle regarding variety registration and other legal processes [Sections 18, 44]. A farmer shall not be liable to pay any fee in any proceeding before the Authority or Registrar or the Tribunal or the High Court under this Act or the rules made there under. "Fee for any proceeding" includes any fee payable for inspection of any document or for obtaining a copy of any decision or order or document under this Act or the rules made there under. However, this exemption does not include fee on lawyers privately hired by farmers to represent them at the Tribunal or Appellate Board or Courts.

6.7.8 FARMERS' RIGHT TO RECEIVE COMPENSATION FOR UNDISCLOSED USE OF TRADITIONAL VARIETIES

It was already discussed that at the time of variety registration, a professional breeder has to disclose about the parentage of his variety to be registered. The cause of benefit sharing arises when the registered variety of a professional breeder is either an EDV or the pedigree of the new variety has certain traditional varieties or traditional knowledge sourced from certain regions/communities. In some instance, it may happen that the breeder of a new variety creates gaps about the parentage of the new variety either due to missing information or due to dishonest or unlawful acquisition of parental material of the new variety. Under such circumstances, if such parental varieties belonged to one or more rural communities, they may be denied the opportunity for benefit share due from the new variety. The communities concerned also may not have the capability to detect such use of their varieties or traditional knowledge in the breeding of a new variety. Under such situations, any third party who has a reasonable knowledge on the possible identity of the traditional varieties or knowledge used in the breeding of the new variety, is eligible to prefer a claim for compensation on behalf of the concerned local or tribal community [Section 41 (1)]. The third party could be an NGO, an individual, a government, or private institution. Such compensation claims are to be submitted to the PPV & FR Authority by such third party. The Authority on verification of the authenticity of the claim shall admit the same and decide on the compensation to be awarded. The awarded compensation will be remitted in the National Gene Fund by the PBR holder (Anonymous, 2006). The National Gene Fund shall disburse the compensation to the party who made the claim.

6.7.9 FARMERS' RIGHT FOR THE SEEDS OF REGISTERED VARIETIES THROUGH COMPULSORY LICENSING

It was mentioned earlier that one of the objectives of the Act is to promote the availability of high-quality seed and planting material to farmers. To restrict the monopoly of professional breeder in terms of their newly developed variety, the PPV & FR Authority creates the scope of granting compulsory license, after 3 years of issue of certificate of registration, in case there is any complain about the availability of the seeds of any registered variety to public at reasonable price or inadequate seed supply [Section 47(1)]. The criteria of licensing would depend on nature of variety, time elapsed since grant of license, price of seed, efforts by the breeder to meet seed requirement of public, and the capacity or ability or technical competence of the applicant to produce and market seed materials.

6.8 NATIONAL GENE FUND

The authority has established National Gene Fund [Section 45] where a share of the royalty earned from these EDVs go to Gene Fund. The breeder who wants to use FV for creating EDVs has to take the permission of the farmers of farm-community. The other sources of the National Gene

Fund as are the annual fee payable to the Authority by way of royalty, the compensation deposit in the Gene Fund, and the contribution from any National and International organization. The National Gene Fund is used for paying any amount as benefit sharing under Subsection 5 of Section 26; for paying compensation under Subsection 3 of Section 41; for supporting conservation and sustainable use of genetic resources, including in situ and ex situ collection; and for strengthening the capability of the panchayat in carrying out such conservation and sustainable use, the expenditure of the schemes related to benefit sharing (Roy, 2013). Therefore, the mandate to establish the National Gene Fund is supporting and rewarding farmers, particularly the tribal, rural communities engaged in conservation, improvement, and preservation of genetic resources of economic plants and their wild relatives particularly in areas identified as agro-biodiversity hot spots. There are provisions for rewarding farmers as individual Farmers' Recognition Award, Plant Genome Saviour Individual Farmers' Reward, and Plant Genome Saviour Community Reward every year for the farmers.

6.9 PROCEDURE FOR REGISTRATION OF FARMERS' VARIETY

An FV can be registered by an individual farmer or a group of farmers or a farming community, provided that the applicant(s) is directly involved in the conservation, improvement, development, and maintenance of the variety. A farming community conserving an FV gains eligibility for its registration only when the farmer or community of farmers who had originally evolved that variety remains unknown (Bala Ravi, 2004). When two or more persons are applicant for any varietal registration, in that case, they enjoy equal ownership of the variety. When a community registers a variety, each member of that community will enjoy equal ownership. A farmers' community may be a group of farming families practicing cultivation or conservation of a crop within a geographic area defined by politico-administrative (panchayat or revenue) demarcation. Equal ownership right within a community means equal right of all the commercial produce, market seed, and license or sell the PBR to other parties. Similarly, the members of the community are equally responsible for safeguarding the PBR (Bala Ravi, 2004). In the case of a variety developed and conserved by a community of farmers, only that community, not its individual members, is eligible to make an application on that variety. The registration of a community variety helps in establishing the legal basis to claim a benefit share whenever that variety is used as a parent to breed a new variety. It gives exclusive right to the community for commercial production and marketing of seed. If the variety is widely cultivated outside the community area, the seed marketing may bring good revenue to the community. The community can also license the PBR to third parties. This exclusive community right on the variety in no way prevents farmers within or outside the community from its cultivation, including seed saving, exchanging, or selling among other farmers.

In this chapter, the details regarding registration of FV will be discussed. First, the farmer or farming community should apply for registration of the variety in the prescribed format of the PPV & FR Act. When the applicant is a community of farmers, the community has to authorize in writing a few of its members as its representatives to file the application on its behalf. All the information on the variety, known to the applicants and as required under the form, has to be accurately provided. There is no need to submit any registration fee for FV. The application should be submitted to the nearest Registrar office of PPV & FR Authority. Every application that is submitted under the Act for an FV, however, must be endorsed by a government agency or official, such as the District Agricultural Officer, Block Development Officer, or District Tribal Officer, or by the Director of Research in a State Agricultural University or research institute, or by reputed NGOs. Such endorsement is intended to serve as an acknowledgement of the authenticity of the community (Kochupillai, 2012). In the second step, the applicant farmer(s) should produce adequate quantity of good quality seed of the variety which the Authority may require for conducting the DUS (distinctiveness, uniformity, stability) test. The requirement of seed is varied based on the crop species, and a part of the seed of the candidate variety must be deposited in the National Gene Bank (NGB) established by the Authority. The acknowledgement receipt issued by the NGB may have to be submitted to the PPV & FR Authority, if so demanded.

An application goes through two stages prior to registration. First, all applications received by the Authority are compiled and published on its website. In the second stage, only those varieties whose applications have been granted certification for "DUS" (distinctiveness, uniformity, stability) testing are subsequently published in the Plant Variety Journal of India, the journal published by the Authority (Kochar, 2010). At this stage, any objections for registration may be put forward for consideration. The details of an application are posted in the Plant Variety Journal of India once it is approved for registration (Lushington, 2012). The specific purpose of such publication is to ensure that there is no opposition to the applicant's claim to the particular variety. Every application requires a processing time, which may range from 8 to 20 months. Registration is granted on the basis of the denomination, also called the "the label or title" of a variety (Lushington, 2012).

6.9.1 DURATION OF REGISTRATION

The period of ownership or duration of a registered plant variety under the PPV & FR Act is short. This period is 18 years for varieties of vines and trees. For other annual crops, the period is 15 years (PPV & FR act, 2001). However, the initial grant of registration is only for a period of 9 years in the case of trees and vines and 6 years for rest of the plants allowed under this Act. In the case of FVs, before the expiry of the said period, the PBR holders have to apply to the Authority for the extension of initial grant period up to 18 or 15 years, as the case may be, in Form PV-6. The PBR automatically lapses on the completion of the duration starting from the day of registration. Once a variety is registered, it is again advertised in the Plant Variety Journal of India as a registered variety.

6.9.2 CRITERIA FOR REGISTRATION

Regarding registration procedure of FV, there is no requirement of Novelty or newness. But the variety must be distinguishable for at least one essential characteristic from any other variety (Distinctiveness) whose existence is a matter of common knowledge in any country at the time of filling the application. This means the new variety that is a candidate for protection should be distinct from all other known varieties including those landraces and traditional varieties, as well as commercialized or protected varieties. Establishing the distinctiveness of a variety requires that it be sufficiently uniform in its relevant characteristics to enable a variety description may be prepared which will distinguish the variety from other varieties of the same species. Thus, to assess Distinctiveness, the characteristics and their states as given in the *Table of Characteristics* in the "Guidelines for Conduct of Test for Distinctiveness, Uniformity and Stability" published by the PPV & FR Authority, Government of India for individual crop can be referred. The variety must be uniform in appearance under the specified environment of its expression (Uniformity) and also must be stable in appearance and its clonal characteristics over successive generations under the specified environment (Stability). Application must be with respect to a variety and need not be accompanied with GURT affidavit (Genetic Use Restriction Technology like "terminator gene"), and passport data of parental lines. For FV, DUS test will be conducted for one crop season at two locations. The uniformity levels for FV for the respective species shall not exceed double the number of off-types such as specified in the "Plant Variety Journal of India" published by the PPV & FR Authority (Figure 6.1).

The design of the tests should be such that plants or parts of plants may be removed for measurement or counting without prejudice to the observations which must be made up to the end of the growing cycle. The

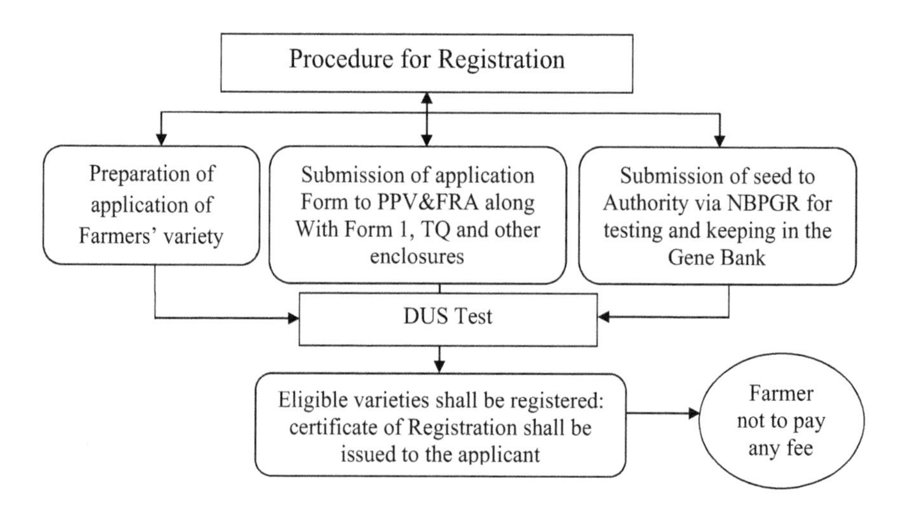

FIGURE 6.1 Registration procedure of farmers' variety.

test plot will be surrounded by one guard row. Additional test protocol for special purpose shall be established by the PPV & FR Authority

6.9.3 CROP SPECIES SPECIFIED FOR REGISTRATION

The Authority has made landmark achievement in the last 10 years. Registration of plant varieties was started by the Authority with 12 crop species in 2007, and now, 144 crop species are covered under the umbrella of the PPV& FR Act for granting IPR (Table 6.1). Nearly, 78 distinctiveness, uniformity and stability (DUS) test centers have been established in State Agriculture Universities (SAU), Indian Council of Agricultural Research (ICAR), Indian Council of Forestry Research and Education (ICFRE), Centre of Scientific and Industrial Research (CSIR), & other reputed research institutes.

6.9.4 STATUS OF REGISTRATION OF FARMERS' VARIETY

About 7527 applications have been received by the PPV & FR Authority (2016) for registration of FV. Amongst these, certificates have been issued only for 765 varieties, and the majority is for rice variety (Table 6.2).

In 2011, in order to accelerate the registration process, experts recommended different methods (or "models") of registration. Two such models are in use: registration initiated by an individual farmer and registration by or on behalf of a farming community by using panchayat-based systems or by means of registered societies (Lushington, 2012). There is no specific requirement that a farmers' community has to be registered. There are instances, however, of groups being registered as producers' groups or societies under different Acts. Every application that is submitted under the Act for an FV, however, must be endorsed by a government agency or official, such as the District Agricultural Officer, Block Development Officer, or District Tribal Officer, or by the Director of Research in a State Agricultural University or research institute, or by reputed NGOs. Such endorsement is intended to serve as an acknowledgement of the authenticity of the community.

The Authority started the registration procedure in 2007 initially for 14 crop species. The highest number of applications was received for rice

TABLE 6.1 The Following Crops Species Have Been Specified for Registration by PPV and FR Authority

Group	No.	Crop species
Cereals	21	Bread wheat, Rice, Pearl millet, Sorghum, Maize, Durum wheat,
		Dicoccum wheat, Other *Triticum* species, Barley, Finger millet, Foxtail millet, two species of buckwheat, Barnyard millet, Four Species of Grain Amaranth Amaranthus hypocondricus, A. cruentus, A caudatus and A. Edulis, kodo millet, little millet, proso millet
Legumes	11	Chickpea, Mungbean, Urdbean, Field pea, Rajma, Lentil,
		Pigeon pea, Black gram, Kidney bean, Soybean, Faba bean
Fibre Crops	06	Diploid cotton (two species), Tetraploid cotton (two species) and Jute (two species)
Oilseeds	10	Indian mustard, Karan rai, Rapeseed (karan sarson), Gobhi sarson, Groundnut, Sunflower, Safflower, Castor, Sesame, Linseed
Sugar Crops	01	Sugarcane
Vegetables and tubers	23	Tomato, Brinjal, Okra, Cauliflower, Cabbage, Potato, Bell pepper, Bitter gourd, Bottle gourd, Chilli, Onion, Cucumber, Pumpkin, Ridge Gourd, Spinach beet, Vegetable Amaranth, elephant foot yam, two species of taro, Paprika,Garlic, Ginger and Turmeric
Flowers and ornamentals	11	Rose, Chrysanthemum, Orchid, Bougainvillea, Canna, Carnation, Casuriana, Chrysanthemum, Gladiolus, Jasmine, Marigold
Spices	05	Black pepper, Coriandor, Fenugreek, Nutmeg, Small cardamom
Fruits	24	Mango, Litchi, Guava, Acid lime, Apple, Apricot, Bael, Banana, Cherry, Custard apple, Grape, Indian jujube (Ber), Jamun, Mandarine, Papaya, Peach, Pear, Pomegranet, Sweet orange, Walnut, Water melon, Aonla, Guava, Mulberry
Medicinal & Aromatic plants	07	Isabgol, Menthol mint, Damask Rose, Periwinkle, Eucalyptus, Brahmi, Neem,
Plantation Crops	02	Coconut, Tea, Beetelvine
Tree crops	03	Karanj, Deodar, Chinar pine,
Others	01	Jatropa

Source: Compiled from data available at http://plantauthority.gov.in/pdf/far-mer.pdf

TABLE 6.2 List of Certificates Issued to Farmers' Variety (up to March 31, 2016)

Sl No.	Crop	No. of Farmers' Varieties registered
1	Rice	749
2	Wheat	03
3	Sorghum	03
4	Chick pea	01
5	Indian Mustard	01
6	Pigeon pea	03
7	Maize	05

Source: Compiled from data available at http://plantauthority.gov.in/pdf/far-mer.pdf.

(4665) followed by maize (230) and mango (186). The initial period of registration, between 2008 and 2010, was marked by slow progress and a subsequent decline in terms of filing (Table 6.3). This may be attributed to various factors. The lack of information among farmers has been cited as a primary cause. Most farmers and farmers' communities lack the knowledge or resources to pursue the various steps of the registration process. and the complexity of filling the registration is also the cause for slower progress of FV registration (Lushington, 2012). During 2009, registration was granted to three FVs of rice from Uttarakhand, namely Tilak Chandan, Hansraj, and Indrasan, which were filled by individual farmers. It was only with the increased involvement of various facilitating organizations that the number of applications increased gradually from 2013 onwards. The highest number of registration was granted during 2014. This increased momentum was achieved mainly due to the large number of applications filed from the state of Odisha. Among the various states, the highest number of applications was received from Odisha, and all together, 694 registrations were granted consisting of individual and community claim. All the applications from this state were under rice. Applications for rice varieties from West Bengal (35) form the next largest group from a State, followed by Kerala (10). In the case of West Bengal and Kerala, all the applications were filled by farmers' society. Most other States have contributed only marginally in terms of the registration process. There appears to be an overall lack of motivation in terms of pushing forward the issue of registration in many States. Organizations

TABLE 6.3 Progress of Applications for Registration of Farmers' Varieties

Sl No.	Year	Total no. of farmers' varieties registered	Detail of farmers' varieties		
			Crop	**No of farmers' varieties Registered**	**State**
1	2009	03	Rice	03	Uttarakhand
2	2012	03	Rice	01	Maharashtra
			Wheat	01	Uttar Pradesh
				01	Punjab
3	2013	45	Rice	26	Odisha
				11	West Bengal
				06	Tamil Nadu
				01	Punjab
				01	Himachal Pradesh
4	2014	459	Rice	456	Odisha
				01	West Bengal
			Chickpea	01	Maharashtra
			Mustard	01	Rajasthan
5	2015	221	Rice	178	Odisha
				13	West Bengal
				10	Kerala
				06	Manipur
				02	Assam
			Sorghum	03	Madhya Pradesh
			Pigen pea	01	Rajasthan
				01	Uttar Pradesh
				01	Madhya Pradesh
			Wheat	01	Rajasthan
			Maize	05	Rajasthan
6	2016 (31st March)	47	Rice	34	Odisha
				10	West Bengal
				02	Tamil Nadu
				01	Manipur

Source: Compiled from data available at http://plantauthority.gov.in/pdf/far-mer.pdf

such as the M. S. Swaminathan Research Foundation (Chennai), Basudha, VRIHII (West Bengal), Biodiversity Management Society (Madhya Pradesh), AMAPCON (Manipur), Seed Care (Kerala), and various State Agricultural Universities have been credited with facilitating the registration process on behalf of farmers. The rural population mainly received their education in the local language. This makes it more difficult to the farmers to read and understand the Act. Training and awareness programs in the local language through extension workers, university teachers, local educated persons, etc. are essential to educate the farmers about the Act, registration process, and application filling. The PPV & FR Authority has organized 1278 programs for awareness and capacity building of different stakeholders and has published literature in vernacular language for effective communication with farmers of States (Anonymous, 2016). Further, no notification of registration of varieties had been published in local newspapers, although farmers are supposed to be encouraged to make such announcements in the print media. Additionally, large farmers may be able to prove the origin of their innovations, while small farmers generally lack the resources to do so. In addition, although various tribal communities have been engaged in preserving traditional knowledge and farming practices, they have received little assistance in terms of the registration process.

6.10 THE PLANT GENOME SAVIOUR COMMUNITY RECOGNITION AWARD

The contribution of large number of individual farmers and farming communities who are engaged for conserving the agro-biodiversity and contributing in the development of many improved varieties goes unnoticed. According to the provisions of section 70 (2) (a) of the PPV & FR Rules, 2003, the Government of India in consultation with the Authority instituted the Plant Genome Saviour Community Award since 2007–2008. Gram panchayats, state agricultural university(s), Krishi Vigyan Kendra(s), Indian Council of Agricultural Research (ICAR) centers, reputed research institutes, non-governmental organization(s), and community-based organizations and farmer's associations can sponsor applications. The award consists of Rs. 10 lakhs in cash along with a citation and a memento. A maximum of 5 such awards are given every year. The selection of awardee(s) is done by a national level committee. The award is applicable to all Indian farming communities, particularly the tribal and rural communities. The applicants are required to submit evidence in support of the conservation work done by them, provide seeds or planting material of the conserved varieties, brief proposal for the utilization of the award money in community welfare, and to inform if the material has been utilized by any breeder in developing any other improved varieties. The application form can be obtained from the office of the PPV&FR Authority, SAUs, ICAR Institute, etc. or can be downloaded from the website www.plantauthority.gov.in.

Plant Genome Saviour Recognition Certificates have been given since 2007–2008 and farmers' communities from 5 states comprising Uttarakhand, Kerala, Odisha, Rajasthan, and Karnataka were the recipients for conservation of folk and traditional landraces of rice, low gluten wheat, pearl millet, chickpea, cumin, and white-seeded horse gram. During 2008–2009, certificates were given to the farming communities from West Bengal, Jharkhand, and Kerala mainly for the conservation of rice landraces having the potential of resistance against bacterial leaf blight (BLB) and drought tolerance and for having unique medicinal properties. During 2009–2010, seven farming communities from Gujrat, Maharashtra, Uttar Pradesh, Madhya Pradesh, Kerala, and Tamil Nadu received this prestigious award. During 2011–2012, 15 individual farmers from nine states were the recipients, whereas during 2012–2013, four farmers from Kerala (03) and West Bengal (01) and during 2013–2014, 11 farmers from Kerala, Chhattisgarh, Goa, and Karnataka were the recipients of the recognition certificate (http://plantauthority.gov.in/PGSFR.htm). With regard to farmer rewards, 10 farmers from Rajasthan, Madhya Pradesh, Uttar Pradesh, Manipur, West Bengal, Karnataka, and Kerala were the recipients during 2012. During 2013 and 2014, 10 and 11 farmers, respectively, from different states like Kerala, Chattisgarh, Karnataka, West Bengal, Manipur, Uttarakhand, Goa, and Pondicherry received the reward.

6.11 IMPLICATION OF THE ACT TO THE FARMING COMMUNITY

The Indian farmers are highly heterogeneous in terms of social and economic status, literacy level, and progressiveness; therefore, the impact of farmers' right will not be uniform in all the cases. The tribal and marginal farmers rarely adopt any new technologies. Therefore, they remain unaffected with any new technologies or amendments in the arena of plant breeding. Hence, government and NGOs should recognize their effort for conservation and maintenance of biodiversity despite their ignorance about new techniques and laws. Small and marginal farmers are resource-poor and lack in technical awareness; they are the most vulnerable section of the farming community to the risk of input farming (Chandrasekharan and Vasudev, 2002). However, if technology dissemination is proper, then this section of farmers can try to adopt new technology and can benefit immensely. The large and progressive farmers are the risk-taking sector of the farming community and take the lead role for adoption and dissemination of any new technologies and act as a bridge between the research institutes, SAUs, and other sections of the farming community. These groups of farmers are financially secure and aware of new technologies, laws, and amendments; therefore, they are immediate beneficiaries or gainer from Farmers' right and variety protection.

The PPV & FR Act will boost plant breeding activities, especially encouraging the participation of private seed industry player beside the public-sector institutes like ICAR and SAUs, resulting in an increased choice of germplasms. However, all the farmers will not get equal amount of benefit because the private seed companies are more concentrated on producing hybrid as one of the component of high input farming; thus, only large and progressive farmers are their major targets. On the contrary, small and marginal farmers still largely depend on traditional landraces and public-sector plant breeding institutes. With regard to registration of FV, large farmers may be able to prove the origin of their innovations; small farmers generally lack the resources to do so. The requirement that innovators ensure that varieties conform to DUS criteria is difficult for small farmers to fulfill. It is clear that farmers need facilitators to work on their behalf, and public-sector institutions have a vital role to play in this regard (Lushington, 2012).

6.12 CONCLUSION

India's PPV & FR Act not only upholds farmers' rights to save, use, and exchange seeds and propagating material but also attempts to enable farmers to claim special forms of intellectual property rights over their varieties. Legal and economic costs of establishing the system, the difficulties of legally claiming rights for farmers, and the limited returns from plant variety protection itself are some of the reasons why IPR-based farmers' rights approaches are unlikely to provide significant economic returns to farmers. Being a pioneer and role model in the protection of farmers' rights, India is duty bound to closely monitor the effectiveness of this regime, not only for benefit of its own farming community, but also for the benefit of the farming communities in other developing countries. In addition to this, in the light of the discouraging trends of registration for FV, efforts must be strengthened to educate farmers about their right under the Act. The issue of concern for developing countries like India rests on how best to harmonize the new system with the interests of economically and socially vulnerable groups and classes, particularly among the working people of rural India and farming communities.

KEYWORDS

- **plant variety protection**
- **PPV & FRA**
- **Rights to Farmer**
- **UPOV**

REFERENCES

Bala Ravi, S., (2004). *Manual on Farmers' Rights*. MS Swaminathan Research Foundation, Chennai, India, 1–80.

Brahmi, P., Saxena, S., & Dhilon, B. S., (2004). The protection of plant varieties and farmers' rights act of India. *Current Sci.*, *86*(3), 392–398.

Chandrasekharan, S., & Vasudev, S., (2002). The Indian plant variety protection act. *J. Intellectual Property Right*, *7*(11), 506–515.

Convention on Biological Diversity, (1993). www.cbd.int/convention/convention.shtml.

Eighteenth Report of the Standing Committee on Agriculture. (2006). Lok Sabha Secretariat.

Kanwar, S., & Evenson, R., (2003). *Does Intellectual Property Protection Spur Technological Change?* Oxford Economic Papers, *55*(2), 235–64.

Kochhar, S., (2008). Institutions and capacity building for the evolution of intellectual property rights Regime in India: III –Conformity and enforcement issue. *J. Intellectual Property Rights*, *13*(5), 239–244.

Kochhar, S., (2010). How effective is sui generis plant variety protection in India? Some initial feedback. *J. Intellectual Property Rights*, *13*(5), 273–84.

Kochupillai, M., (2011). The Indian PPV & FR act: Historical and implementation perspective. *J. Intellectual Property Right*, *16*(3), 88–101.

Kumar, P. S., Khan, S. M., Hora, M., & Rao, M. P., (2011). Implementation of Indian PPV & FR act and rules: Inadequacies leading to avoidable litigation. *J. Intellectual Property Right*, *16*(3), 102–106.

Lushington, K., (2012). The registration of plant varieties by farmers in India: A status report review of agrarian studies. *Review Agrarian Studies*, *2*(1), 112–128.

Nagarajan, S., (2010). Paper presented on National Consultative Seminar on Section 41 of PPV & FR Act, 2001: Rights of Communities, New Delhi, 25–26 May. http://www.plantauthority.gov.in/pdf/Proceed_rights of community.pdf (December 4, 2010).

Plant Variety Journal of India, various issues. Available at http://plantauthority.gov.in/publications.htm.

Ramanna, A., (2006). *Farmers' Rights in India*, FNI report 6/2006, Farmers' Rights Project, Lysaker.

Roy, B., (2013). Farmers' rights and its justification under Indian circumstance. *J. Crop Weed, 9*(1), 139–144.

Sahay, B. R., & Srivastava, M. P., (2001). *National Agricultural Policy in the New Millennium*. Anmol Publications Pvt. Ltd., New Delhi, 1–14.

Seema, P. S., (2012). *Protection of Farmers' Rights In India Challenges for Law in the Context of Plant Breeders' Rights.* Cochin University of Science and Technology Cochin-682 022.

Srinivas, C. S., (2003). Exploring the feasibility of farmers' rights, development policy review. *J. Agricultural and Development Economics*, *21*(4), 419–47.

Srinivasan, C., (2005). International trends in plant variety protection. *J. Agricultural and Development Economics*, *2*(2), 182–220.

The Patent act, (1970). www.wipo.int/edocs/lexdocs/laws/en/in/in065en.pdf.

The Protection of Plant Variety and Farmers' Rights Act, (2001). Universal Law Publishing Co. Pvt. Ltd., C-FF-1A, Dilkhush Industrial Estate, Delhi – 110 033.

TRIPs Agreement, (1994). https://www.wto.org/english/docs_e/legal_e/27-trips.pdf.

Wood, D., & Jillian, M. L., (1997). The conservation of agrobiodiversity on farm: Questioning the emerging paradigm. *Biodiversity Conservation*, *6*, 109–129.

CHAPTER 7

PATENTING OF MICROBIOLOGICAL AND BIOTECHNOLOGICAL INVENTIONS: THE GLOBAL AND INDIAN SCENARIOS

RENU, ABHISHEK PARASHAR, SANJAY KUMAR GUPTA, PRAMOD KUMAR SAHU, UPASANA SAHU, HARDESH KUMAR, KHAN MOHAMMAD SARIM, ARVIND GUPTA, and PAWAN KUMAR SHARMA

ICAR-National Bureau of Agriculturally Important Microorganisms, Kushmaur, Maunath Bhanjan–275103, Uttar Pradesh, India, E-mail: renuiari@rediffmail.com

CONTENTS

ABSTRACT

Scientific inventions in the field of microbiology and biotechnology are the most happening in today's world. There has also been great awareness in protection and promotion of economic and commercial interests. Besides, there has been a quantum jump in research and development (R&D) costs in various sectors, including biotechnology and microbiology, along with an associated jump in investments required for commercializing a new technology. Hence, the developers of technology require high stakes, and in turn, the need to protect the knowledge from unlawful use has become expedient, at least for a period, that would ensure recovery of the R&D and other associated costs and adequate profits for continuous investments in R&D. Intellectual property rights (IPR) is a strong tool to protect investments, time, money, and effort invested by the inventor/creator of an IP, because it grants the inventor/creator an exclusive right for a certain period of time for use of his invention/creation. Thus, IPR helps in strengthening the economy of a country by promoting healthy competition and encouraging industrial development and economic growth.

7.1 INTELLECTUAL PROPERTY RIGHTS: AN INTRODUCTION

IPR has been defined as ideas, inventions, and creative expressions based on which there is a public willingness to bestow the status of property. IPR is a right given by the government to inventors for their intellectual work in order to enable them to reap commercial benefits from their creative efforts or reputation. IPR is given for stipulated period of time after disclosure of work in public domain.

There are several types of intellectual property (IP) protection, and Article 2(viii) of the convention establishing the World Intellectual Property Organization (WIPO) provides that "intellectual property shall include rights related to:

- literary, artistic, and scientific works;
- performances of performing artists, phonograms, and broadcasts;
- inventions in all fields of human endeavor;
- scientific discoveries;
- industrial designs;
- trademarks, service marks, and commercial names and designations;

- protection against unfair competition; and
- all other rights resulting from intellectual activity in the industrial, scientific, literary, or artistic fields.

7.1.1 NEED OF IPR

7.1.1.1 Patent Protection

Biological inventions can be protected under patent. Any work that has novelty and industrial application can be considered for grant of patent. Companies can protect their costly but commercially viable work through patent; further, it prevents unauthorized use of invented work.

7.1.1.2 Revenue Generation

Inventions with commercial potential are a big source of revenue generation. Genetically modified seed to vaccine making has given companies immense profit. Further, due to patent, work can be transferred to other companies on the basis of license agreement for monetary benefit.

7.1.1.3 Reward

IPR helps in rewarding those people who have done the hard work and encouraging people to do more research that are commercially viable. This kind of encouragement is even more important if funding is given by the state government. Most of the claim in biotechnology invention is related to gene or protein sequence, plant and animal tissue culture, monoclonal antibody, diagnostic kit, microorganisms, and vaccine composition in the form of product or process. Oil-degrading bacteria, Harvard mouse, and herbicide-resistant plants are some of the examples that are commercially in use.

7.2 HISTORICAL DEVELOPMENT IN INTELLECTUAL PROPERTY PROTECTION REGIME (WIPO, WTO, AND TRIPS)

The majority of countries in the world are having their own IP protection and enforcement system as it encourages innovation and creativity, which in turn leads to economic prosperity of the nation. The first IP law that protected the investors' interest against copying of their creation was passed in Venice in 1474. This was soon followed by England, wherein in 1624, the government passed the Statute of Monopolies, which granted IPRs to the inventor for a limited period. The need for a system to protect IP internationally arose when foreign exhibitors refused to attend an international exhibition of inventions in Vienna in 1873 because they were afraid that their ideas would be stolen and exploited commercially in other countries. This gave birth to the Paris Convention for the Protection of Industrial Property in 1883, which made it easier for individuals in one nation to obtain protection globally in form of industrial property rights. This convention was followed by the Berne Convention for the Protection of Literary and Artistic Works through which the copyright entered the international arena in 1886. This convention was aimed to help nationals of its Member States obtain international protection of their right to control, and receive payment for, the use of literary and artistic works. International bureaus were set up for both the conventions to carry out administrative tasks, such as organizing meetings of the Member States. These bureaus united in 1893 to form an international organization called the United International Bureaus for the Protection of Intellectual Property (BIRPI, a French acronym). BIRPI, which was based in Berne, Switzerland, with a staff of seven, became predecessor of what is today known as WIPO. WIPO is a specialized agency of the UN, with a mandate to administer IP matters recognized by the UN Member States. WIPO looks after about 21 international treaties in the field of IP that are broadly divide into three groups, namely treaties that establish international protection; treaties that facilitate international protection, and treaties that establish classification systems.

The extent of IP protection and enforcement varied widely around the world, and these differences were a source of tension in international economic relations. There was thus a need for harmonization and predictability for disputes to be settled more systematically. After the end of World War II, in 1948, General Agreement on Tariffs and Trade (GATT) was established for the regulation of international trade. In 1986, Uruguay round was started for trade-related issues. Its tenure was of 87 months. There were several trade-related subjects that were discussed in this round, such as subsidy in agriculture, textile, creation of World Trade Organization (WTO), and role of IPR in trade. After the end of Uruguay round, in

Marrakesh, WTO was created in place of GATT on January 1, 1995. WTO headquarter is in Geneva, and its main role is to administratively control international trade. Like GATT, it regulates international trade, giving the platform to negotiate and settle down any trade-related disputes between two countries. It is also responsible for making new trade laws on the basis of its ratification done by Member Countries of WTO.

The WTO Agreement on Trade-Related Aspects of Intellectual Property Rights (TRIPS Agreement), which came into force in 1995, brought with it a new era in the multilateral protection and enforcement of IP rights. The Member Countries under this were given time till January 1, 1996 to ratify it for the protection of IPR. The idea was to protect inventor's work not only in his or her native country but also in Member Country of WTO. The agreement includes general provision, basic principles, and final provision under part I and VII. Under this section, IPR has been defined to include patents, copyrights, trademarks, plant variety protection, trade secrets, industrial designs, and geographical indicators. This section also says that each Member Country will ratify the TRIPS agreements and must give the same treatment whether the persons or organization or company is from native country or Member Country if protection has been filed against IPR-related work in that particular country. Part II involves standards concerning the availability, scope, and use of IPR. This section gives information of various rules and regulations that will be made by Member Country in the area of IP so that they can fulfill the agreement that has been made in TRIPS. For example, patent is only given to work that has inventiveness and industrial application (Article 27.1). According to Article 27.2, no patent will be given to work that involves protecting humans, plants, or animals on safety ground or morality issue. In Article 27.3, patents is also not given to essential processes involved in reproduction of plants and animals or method of diagnostic, therapeutic, and surgical process that is used for the treatment of animals and humans. However in the *sui generis* system, plant variety can get protection, according to Article 27.3(b). To prevent misuse of patent, exclusive marketing right of patent is given (Article 28.1); further, a patent is granted for 20 years (Article 33). Part V of agreement deals with prevention of disputes and settlement related to IP rights. TRIPS also make arrangements for coordination between WTO and WIPO for better implementation of IP rights at the worldwide level.

Provisions in the TRIPS Agreement concerning copyright and related rights, patents, trademarks, geographical indications, industrial designs, and layout designs of integrated circuits, directly complement the international treaties administered by the WIPO secretariat. An agreement between WIPO and the WTO since 1996 provides for cooperation concerning the implementation of the TRIPS Agreement, such as notification of laws and regulations and legislative assistance to Member Countries.

IP is divided into two categories: industrial property, which includes patents for inventions, trademarks, industrial designs, and geographical indications, and copyright and related rights, which cover literary and artistic expressions (e.g., books, films, music, architecture, art), plus the rights of performing artists in their performances, producers of phonograms in their recordings, and broadcasters in their radio and television broadcasts, which are also referred to as neighboring rights.

7.3 TRENDS AND ADVANCES IN MICROBIOLOGY AND BIOTECHNOLOGY

Traditional microbiology and biotechnology were largely confined to two major areas: breeding and industrial microbiology. But, with the advent of genetic engineering and sequencing technologies, there has been a revolution in these research fields, and the whole complexity of life forms, especially that of microbes, has changed. Recombinant DNA technologies have been employed in the production of genetically modified organisms with altered genetic constitution such as overexpression of desired proteins, value addition, and much more. Advent of various sequencing platforms have given insight into various gene function and discovery of new genes, thus paving way to better understand functioning of an organism. The high-ended techniques like metagenomics, metatranscriptomics, and metaproteiomics have enabled to isolate and identify the genes and their functions without isolating microorganisms.

7.4 PATENT REGIME IN MICROBIOLOGY

Patent is a recognition for an invention, which satisfies the criteria of global novelty, nonobviousness, and industrial application. IPR is a prerequisite for better identification, planning, commercialization, rendering, and thereby protection of invention or creativity. Each industry should evolve its own IPR policies, management style, strategies, and so on depending on its area of specialty. Biotechnology and microbiology industry currently has an evolving IPR strategy that requires a better focus and approach in the coming era.

7.5 THE GLOBAL SCENARIO

7.5.1 PATENTING OF LIVE ORGANISMS

An idea that is tangible, nonobvious, novel, and has utility can come under patent criteria. Patenting of a live organism is always a highly controversial issue, and whether microorganisms fall under this category has been a debatable issue across the globe.

Article 27(b), Para 3 of TRIPS states that the following cannot be excluded from patentability: (i) microorganisms, e.g., bacteria, viruses, fungi, algae, protozoa, etc., and (ii) nonbiological and microbiological processes of production of plants and animals, wherein the Member will provide protection to plant varieties either by patents or by an effective *sui generis* system or any combination thereof .

The patenting of life forms have always been a point of concern on which the industrially developed nations thrived upon. Various milestones in patenting life forms in the IPR regime are shown in Table 7.1.

There are different laws in developed countries for patenting microorganisms or live organisms.

7.5.2 UNITED STATES

According to the American law, patent can be granted to nonnatural, nonhuman multicellular organisms created by human intervention under US patent law 35 USC 101. A famous case of Diamond vs. Chakrabarty occurred in 1970 in which the United States Patent Office (USPTO) gave no patent on oil-degrading bacteria developed by Dr. Ananda Mohan Chakrabarty. *Pseudomonas* bacteria, which was genetically modified by Chakrabarty to degrade complex organic compounds to simpler ones, thus

TABLE 7.1 Timeline for Milestones of Patents Granted on Microorganisms

Year	Breif description of patent
1873	Pasteur got US Pat no 141.072 for Yeast an article of manufacture
1969	Animal Breeding Methods – German Federal Supreme Court accepts
1975	Microorganisms are patentable – German Federal Court
1977	Budapest Treaty Signed
1980	Microorganisms become patentable in USA (Diamond *vs.* Chakrabarty)
1985	Plants/tissues/ tissue culture, Seeds become patentable: US PTO
1987	Multicellular Organisms are patentable – US PTO
1988	European Patent Office grants first patent on plant – US PTO issues patent on "oncomouse"
1995	DNA not life but chemical and patentable – EPO declaration
1998	First patent covering ESTs by Incyte Pharmaceuticals
2001	Oxford GlycoSciences filed patent for 4000 human genes and proteins and codes for them; India joined Budapest Treaty
2002	Dimminaco AG case
2005	New Patent Regime

reducing pollution caused by oil spills. The work has both novelty and utility for grant of patent. The claim was filed under the title 35 US Code 101, which authorizes patent for only newly made composition of matter. Later, US Supreme Court gave the verdict in favor of Chakrabarty, and oil-degrading bacteria were the first live organism that got the patent (US Patent 4259444 A). In 1988, Harvard mouse (transgenic oncomouse, US Patent 4736866 A) was granted patent as the breast cancer gene was artificially incorporated into mice genome through genetic engineering, thereby making it more susceptible to cancer. The work provided better understanding of cause and progression of breast cancer. Likewise, the American patent law also gave rights to inventors for patenting of transgenic plant such as herbicide-resistant variety. Further, patent was also granted to plants that are propagated in vitro or asexually (US patent law 35 USC 161). Monsanto Company in 2011 received the patent on transgenic plant that contains a recombinant gene G1073 producing a transcription factor, thereby making the plant drought resistant (US Patent 7888557 B2). In 1997, RiceTec received the patent on basmati rice, but as basmati rice is grown long before in India as a food material, appeal was filed against its patent. Ultimately, RiceTec lost the patent of basmati on most of its strains. Myriad Genetics working on cancer research discovered that mutation in BRCA1 & 2 (tumor suppressor protein) makes female more susceptible to breast cancer. On the basis of this assumption, they developed a diagnostic method for the detection of mutation in the BRCA gene and at the same time applied for patent of this gene. The US patent court earlier gave patent on gene that is novel in nature. However, in 2013, Supreme Court gave a decision that mere isolating a gene naturally cannot come for patent grant, although cDNA which is not natural, can be patented. Hence, in the US, if an idea is novel, new, nonobvious and has utility, it receives patent.

7.5.3 EUROPE

Under Section 53 (a), patent is not granted to new plant or animal variety; moreover, essential biological processes that are responsible for plant and animal reproduction cannot received patent. Section 53(b) prevents patent of transgenic animals or plants on the basis of morality. To resolve the issue related to patent of biotechnology-related research, European Union (EU) has adopted biotechnology directive, which is also ratified by the European Patent Office (EPO). Under biotechnology directives, patent can be granted on invention related to microorganisms. The work could be microbiological process or product formation. On the basis of morality, it was decided that no patent will be granted related to cloning of human beings, commercialization of human embryo, or any modification of the genome of an animal that has no medical benefit to humans and further causing the animal to suffer. Under biotechnology directive Articles 8 and 9, patent could be granted to transgenic animals and plants on the basis that human intervention is involved in essential biological processes such as reproduction. In Article 9, gene sequence will receive a patent and at the same time its incorporation to animal will further protect the animal and other animals that reproduce through that parent animal (transgenic animal).

7.5.4 JAPAN

Japan also started to grant patents for inventions related to plants (seeding law), animals, and microorganisms following United States. Article 32 of Japan says that no patent can be granted to any invention that is injurious to health or is against the law or morality. It has been seen that unlike Europe, the morality issue in Japan is less serious related to grant of patent for those inventions that are ethically questionable. Patent can only be granted to any work if it has industrial application according to the law of Japan. In 1997, Japan started to grant patents in the field of biotechnology by making special guidelines under Chapter 2 of the Patent Act. Biotechnological inventions are divided in four groups: genetic engineering, plants, animals, and microorganisms. Genetic engineering invention includes monoclonal antibody, recombinant protein production, gene, vector, recombinant vector-making process, etc. Plants and animals section contains protection of parts of plants or animals and processes involved in the creation of plants and animals. Microorganisms in light of patent here includes yeast, bacteria, actinomycetes, molds, mushrooms, unicellular algae, viruses, protozoa, and further undifferentiated animal or plant cells as well as animal or plant tissue cultures. Patent deals with the protection of work that makes microorganisms useful for humans.

7.5.5 CHINA

China allows claims pertaining to microorganisms. Patenting of DNA sequences is permitted as a large chemical compound or compositions of matter. According to product(s) of nature rule, naturally occurring DNA sequences are unpatentable in China. But, judicial interpretation has resulted in patentability claims that cover purified and isolated DNA sequences as new composition of matter. Hence, in China, the bottom line is purify, isolate, and patent.

7.5.6 AUSTRALIA

The IP Australia Fact Sheet states that standard patents for biological material can be obtained, which include microorganisms, nucleic acids, peptides, and organelles provided that this material has been isolated from its natural environment, or has been synthetically engineered, or is a recombinant. DNA or genes in the human body are not patentable, but if they are isolated from a human body, they may be patentable.

7.5.7 *BUDAPEST TREATY AND CREATION OF THE INTERNATIONAL DEPOSITORY*

The Budapest Treaty on the International Recognition of the Deposit of Microorganisms for the Purposes of Patent Procedure (referred to as the "Treaty") was adopted by the Budapest Diplomatic Conference on April 28, 1977, and it entered into force on August 19, 1980. Regulations were also adopted under the Treaty in the Conference. For the grant of patents, disclosure of the invention is a generally recognized requirement as a quid pro quo to the public, which is done normally through a written description. Such a description is not sufficient for disclosure where an invention involves a microorganism or other biological material (hereinafter referred to as "microorganisms") or the use of it (in particular in agriculture, food, and pharmaceutical industries), which is not available to the public. Hence, in the patent procedure of an increasing number of countries, it is necessary not only to file a written description but also to deposit a sample of the microorganism at a specialized institution. Patent offices cannot handle microorganisms as they are not equipped with the expertise and infrastructure, which is a costly affair. When protection is sought in several countries for an invention involving a microorganism or the use of a microorganism, the complex and costly procedures of the deposit of the microorganism would have to be repeated in each of those countries. The Treaty was concluded in order to eliminate or reduce such multiplication and to enable one deposit to serve the purpose of all the deposits, which would otherwise be necessary. It was decided that instead of depositing the microorganism for patent procedure in countries that are part of the Budapest Treaty, it will be better to deposit microorganisms only to an International Depository Authority (IDA) that may be present at inventor's home country or outside. Today, 80 countries are part of this treaty, and any country may get its membership; however, it has to first ratify Paris convention on protection related to the industry.

7.6 THE INDIAN SCENARIO

The Indian Patent Act was passed in 1970. The Act was amended in 1999, 2002, and 2005 according to the requirement to meet the TRIPS agreements. Before 2002, India did not grant any patent in the field of biotechnology or to any invention related to live organisms. Later, in the second amendment of the Patent Act, which was passed in 2002, it was decided that biotechnological or microbiological processes will be considered as chemical processes, and patent will be given in this field. In 2001, India ratified the Budapest Treaty for patenting and deposition of microorganism-related work. National Center for Cell Science (NCCS), Pune, India, and Microbial Type Culture Collection and Gene Bank (MTCC) at IMTECH, Chandigarh, India, are recognized as IDA where microorganisms such as bacteria, plasmids isolated from bacteria, yeast, bacteriophages, and fungi can be deposited.

India also passed the law on conservation of genetic resources of biological materials (Biological Diversity Act, 2002). In Section 10 of the Patent Act, an amendment was made in 2005 that inventor must disclose the source and geographical location of the biological material if he or she has used it for invention. Prior approval of National Biodiversity Authority is required for accessing biological material by foreigners/NRIs (Section 3) as well as before seeking patent based on biological materials and traditional knowledge (TK) obtained from India (Section 6(1) of Biodiversity Act 2002). Under Section 3j and 3i of the Patent Act, a patent cannot be granted to animal or plant variety, seeds, and essential biological processes involved in the reproduction of animals or plants. Further, no patent is granted for the processes used in medical, surgical, or curative treatment related to humans or animals and methods involved in increasing the economic value of animals. This is accordance with TRIPS agreement Article 27.3 (a) & (b). For plant variety protection, India has adopted the *sui generis* system, as plants are not granted patents in India. In this system, the plant variety can be protected by breeders (cultivators or person who has developed or discovered the variety). This gives farmers the right to protect the plant variety. The plant variety must fulfil certain criteria to receive recognition as a new variety, such as distinctiveness, uniformity, stability, and novelty in terms of commercial use.

7.6.1 PATENT ACT AMENDMENTS

7.6.1.1 First Amendment in 1999

- Exclusive marketing right (EMR): The right creates monopoly and gives commercial benefit to the inventor. It was decided that any application for patent that has been filed on or after January 1, 1995 in the convention country and has been given grant in that country on or before January 1, 1995 will get the exclusive marketing right in India if the same work is filed for patent in India. The EMR is given for 5 years, and Section 24A and 24B of the Patent Act deals with this right.

- Mail box facility for pharmaceutical and agrochemical industry under Section 5(2) of the Patent Act. This enables the Member Country to know the name of company or the individual who has been granted the patent for a particular work in that country.
- Compulsory licensing: It works opposite to EMR. In EMR, the inventor creates monopoly, but in compulsory licensing, the inventor cannot negotiate his or her work for commercial use. Here, a lump-sum of money is given which is far below the price decided during EMR. Compulsory licensing is given only after three years of grant of patent and is given when the said company having EMR for that patent is not able to fulfil the requirement of public or selling the product at a much higher price than its manufacturing cost. Section 24C and 24D deals with this licensing. All in all, compulsory licensing reduces the price of the product and reduces the effect of EMR.

7.6.1.2 Second Amendment in 2002

The following are the salient points considered in the second amendment in 2002:

- It was decided that patent will be granted for 20 years.
- National Biological Diversity Act was passed in 2002 for the conservation of biological material.
- Any invention that is related to national security and comes in for national interest need not to be disclosed under Section 39.
- IPR appellate board was established to deal with the disposal of patent appeal for its fast prosecution.
- No patent will be granted on whole or in parts of animals and plants; further, patenting of biological process or invention related to TK cannot come under patent criteria. However, for microorganisms,

patent can be granted if its function is modified by human intervention and has utility. This comes under Section 3(j).

- The definition of invention has also been changed, and invention is now related to any process or new product that has an inventive step and utility, as previously it meant the manner of manufacturing (Section 2j).
- Under Section 10 of the Patent Act, it was decided that patent will be granted in biotechnology field, and microorganisms could be deposited for patent purpose.

7.6.1.3 Third Amendment in 2005

- The Budapest Treaty was ratified for patenting of microorganisms. Microorganisms can be deposited at IDA. Section 2(1) was amended for this purpose.
- Under Section 3, patent cannot be provided to any substance whose new form has been identified or created unless it has enhanced efficacy. Supreme Court on that basis rejected the patent of gleevec (anti-blood cancer drug) by Novartis in India despite that it has enhanced physical efficacy as compared to the older version of the drug. According to the court, the new version has no known therapeutic efficacy, or in other words, cancer patients receive the same benefit from the drug regardless of whether it is in older form or new form when given.
- In Section 7(3), the word "person" is changed to "owner" in claiming patent of a particular work.
- Under Section 9(1), if the application for patent is filed in provisional specification, it must be filed with complete specification within 12 months from the date of provision application, else it will be rejected.
- Under Section 11, it was decided that the patent office will only examine the application when the request will be made by the applicant in a prescribe format.
- Under Section 10, each inventor should provide information of source and geographical location of the living organism if it was used for the inventive work.

7.6.2 CONDITIONS FOR GRANT OF PATENT

7.6.2.1 Novelty

To receive the patent for a particular work, it should never be disclosed in public domain prior to it its submission to the patent office. The inventor should search for prior art to see whether identical or similar work is available. Prior art search can be done at the website of the EPO, USPTO, or Japanese patent office (JPO). It has made patent prosecution process very fast; further it prevents the duplication of work, saves time and money of the inventor if he or she knows that the said work has already been done, gives information of competitors' work, trends in technological development, etc.

7.6.2.2 Inventiveness

It means that some skill should be involved in the work.

7.6.2.3 Usefulness

The work should have utility or commercial application.

7.6.3 NONPATENTABLE ITEMS

- Any invention that is against the law, morality, or injurious to public such as human cloning.
- Methods applicable in agriculture or horticulture.
- Scientific principles (Newton's law of gravity) or mathematical models.
- Invention in the field of atomic energy.
- Methods involved in medical or surgical treatment of humans or animals.
- Methods enhancing efficiency of machines.
- Invention based on TK cannot be patented.

7.6.4 PROCESS OF PATENT FILING IN INDIA

In India, patent can be filed in all four metro cities, i.e., Chennai, Mumbai, Delhi, and Kolkata. Kolkata is currently the head office for patent application processing. There are four application forms that are used for filing patent application.

7.6.4.1 Form 1

The names of the applicant and inventors are given. The type of application is mentioned, i.e., whether it is a convention or PCT (patent cooperation treaty) or ordinary one. For international patenting, PCT filing is done to protect the invention at the international level. The treaty was signed in Washington in 1978 and India is part of this group. PCT gives a unified way of patent filing for the inventor in as many as countries that are part of PCT and where patent protection is called by the inventor. First, the application is filed in the home country where international search is done by the patent office to see whether the work is novel or not and has commercial application. If the said work is not found in prior art search, a second opinion is taken through international preliminary examination, which is also done by the patent office and is more exhaustive in nature. PCT gives the priority date, which is the application filing date, and within 30 months (In India, it is 31 months) of time from the date of priority, the application can be filed at the national level. At the international level, the application is filed separately to other Member Countries of PCT for seeking patent protection there. The priority date gives an advantage to the inventor by protecting him or her in a situation when other people are also working on the same problem. If an inventor has a priority date for a specific work, other persons could not get the priority date for the same work.

Convention application (Paris convention) is also a kind of patent filing procedure at the international level. This application is filed for the protection of invention related to industrial designs and trademarks. It also gives a priority date. However, here only 12 months are given to the inventors from the date of priority for filing application for patent in member countries that are part of Paris convention. It is important to remember that both application (Convention & PCT application) gives priority date;

however, it is absent in the case of national patent application filing, which is also known as ordinary application.

7.6.4.2 Form 2

Form 2 deals with complete specification where the full descriptive information of work under consideration for patent is given. Complete specification contains- title, field of invention mentioning the subject in which work was carried out, state of art in that field, or in other words, what kind of work has been done in that field previously, objective of invention (problem & solution), detail description of invention with drawing, application of work, and claims. Provisional specification is given when the work in the conceptual stage and there is delay in giving full descriptive information for that work. In any case, complete specification must be given within 12 months from the date of provisional specification filing; otherwise, the application will not be considered for patent.

7.6.4.3 Form 3

This application mentions whether the patent is filed in foreign country or not.

7.6.4.4 Form 5

This application deals with inventorship. The names of inventors and the applicant are mentioned, and they give their approval that they are the true persons responsible for this work.

All four applications along with the requisite fee and a soft copy are send to the patent office for its consideration for patent. A receipt is given by the patent office bearing the application number, place of patent filing, and year of patent filing, for example, 3703 (application no.)/DEL (Delhi)/2013 (year). The patent office publishes the work within 18 months from the date of filing of application. After the publication, pre-opposition could come. In that case, the applicant has to justify the objection, and if the patent office is satisfied, the application is consider for its examination on the request of the applicant. The patent office gives 48 months for patent examination. After its examination, the application can further be objected, which is also known as post-opposition. The patent will be granted after clearance of objection or else rejected. The duration of patent is for 20 years, and renewal is required every year (Figure 7.1).

7.6.5 REVOCATION OF PATENT

- Against the national law,
- Novelty is missing,
- Ownership is wrongly obtained,
- Duplication of work,
- Work is not fully described in the application,
- Using traditional knowledge for invention,
- Secretly commercialization of work before the grant of patent.

In 2013–2014, approximately 42,000 applications were filed for patent, but hardly 4200 were granted. Among the Indian institutes/organiza-

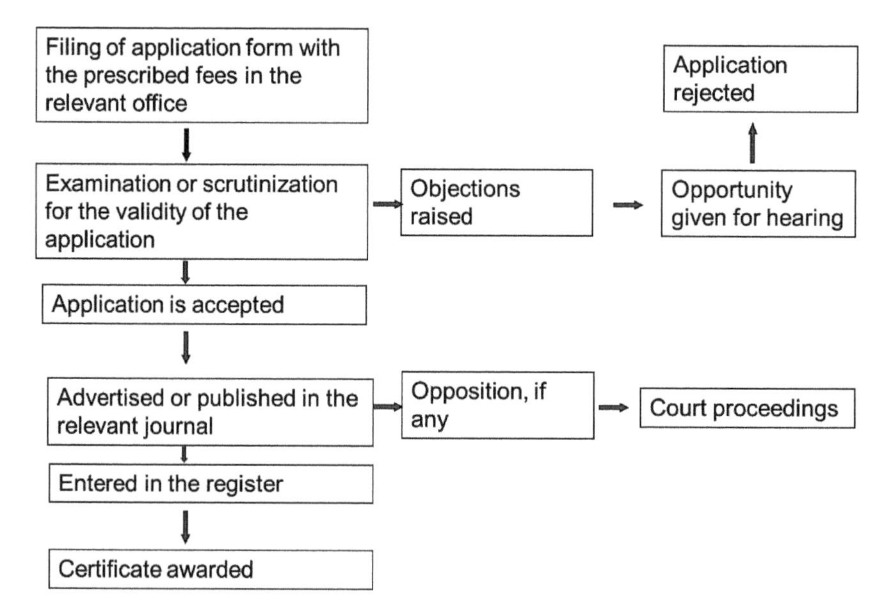

FIGURE 7.1 Diagrammatic representation of the patent filing process.

tions, IIT filed the maximum number of patent application, while CSIR received the maximum number of grant. The Indian patent office also generated 188 crores of rupees as revenue in 2013–2014. Out of 42000 applications, 11000 were filed by Indian applicants, while 32000 were from outside of India. PCT national level patent filling was 27000. USA, Japan, and Germany were the leading countries in filing patent application in India through PCT. The number of PCT international patent application of Indian origin was 816. Out of the 4200 granted patents, 634 were of Indian applicants, while 3592 applicants were of foreign nation. Chemical and related field received the highest number of patent granted (1111 applications), while 220, 51, 44, and 21 applications were from biotechnology, food science, microbiology, and agrochemical field respectively.

7.7 CONCLUSION

In the present era, microbiology and biotechnology need to address key health and agriculture challenges facing poor countries; they also have to assess and manage potential threats to human health and environment. Hence, this subject and field require continuous technological progress and better technological transfer to maintain commercial competitiveness. IPR make it possible to develop strategies for dissemination and transfer of technologies in such a way that may provide maximum social benefits and support the competitiveness of industry.

It is also believed that IPR can create monopoly and could increase the price of essential commodity, thus affecting poor people. Gleevec is one of the examples where Novartis was selling the drug in India at the price of USD 2200 as compared to the Indian company making the same drug in the form of generic drug in USD 100 to 200. Patentability of microorganisms subjected to novelty, nonobviousness, utility, and human interventions paves way to protect microbial wealth from nation's interest perspective, as the microorganisms isolated from various parts of a country like India, and the resultant new, modified, and improved use thereof, should be afforded maximum protection.

All in all, IPR is needed for today's world; it is about promoting commercial aspect of business without harming the interest of the society at large. This can be done by balancing the interests of the inventors & users by making careful law and its effective implementation.

KEYWORDS

- **biotechnological patent**
- **microbes**
- **patent**
- **WIPO**
- **WTO**

REFERENCES

Annual Report, (2013–14), Intellectual property India. www.ipindia.gov.in/cgpdtm/AnnualReport_English_2013_2014.pdf (accessed Aug 10, 2016).

Ashwin, S., (2016). Patenting micro organisms globally http://www.luthra.com/admin/article_images/patenting-micro-organisms-globally.pdf (accessed Aug 10, 2016).

Budapest treaty on the international recognition of the deposit of microorganisms for the purposes of patent procedure note by the Secretariat http://www.wipo.int/export/sites/www/treaties/en/registration/budapest/pdf/wo_inf_12.pdf (accessed Aug 10, 2016).

Chawala, H. S., (2005). Patenting of biological material and biotechnology. *Journal of Intellectual Property Rights*, *10*, 44–51.

Conservation of Bio-diversity Act, 2002 and Biological Diversity Rules, 2004.

Diamond vs Chakrabarty, (1980) 447 U.S. 303.

Dimminaco Ag v Controller of Patents & Designs and Others, [2001] AID No. 1.

Filing of patent application, Intellectual property India www.ipindia.nic.in/ipr/patent/manual/HTML (accessed Aug 10, 2016).

Giugni, D., & Giugni, V., (2013). Intellectual property: a powerful tool to develop biotech research. *Microbial Biotechnology*, *5*, 493–506.

Gosset, N., (2016). Patentability of biotechnological inventions. *eHow Contribution*. www.ehow.com (accessed Aug 10, 2016).

Indian Patent Law, office procedures and examination practice for biotechnology patent applications. Dr. Kardam, K. S., Deputy Controller of Patents & Designs Indian Patent Office. http://www.wipo.int/meetings/en/doc_details.jsp?doc_id=154243 www.wipo.int/edocs/mdocs/sme/en/wipo_ip_del.../wipo_ip_del_10_theme05_2.ppt.

Janice, M. M., (2008). *Biotechnology Patenting in India: Will Bio-Generics Lead a 'Sunrise Industry' to BioInnovation, No. 2*, University of Missouri-Kansas City Law Review.
Jindong, S., Kimberly, Z., Jingrui, W., Changlin, F., Stanton, B. D., & Linda, L., (2013). Transgenic Plants. US Patent 20130191942.
Kulkarni, M., (2016). India, biotechnology and patents: Industry perspective, http://www.bicpu.edu.in/ ipr_ppt/15/kulkarni.pdf (last visited on Sept. 12, 2013) www.fbae.org (accessed Aug 10, 2016).
Kumar, S., (2008). Patentability of biological material(s) – Essentially, therapeutic antibodies – in India, SCRIPT-ed, *5*(3), http://ssrn.com/abstract=1578224 (accessed Aug 10, 2016).
National Biodiversity Authority website. www.nbaindia.org (accessed Aug 10, 2016).
Robert, Cook-Deegan, (2013). Are Human Genes Patentable? *Ann. Intern. Med.*, *159*(4), 298–299.
Shiv, S. S., (2005). *The Law of Intellectual Property Rights*, Deep & Deep Publications Pvt. Ltd., New Delhi, pp. 357.
Sreenivasulu, N. S., & Raju, C. B., (2016). *Biotechnology and Patent Law-Patenting Living Beings* (ebook) 1st edn., 2008 (accessed Aug 10, 2016).
The Patents (Amendment) Act (2005), Intellectual property India, www.ipindia.nic.in/ipr/patent/patent_2005.pdf (accessed Aug 10, 2016).
The Patents Act 1970, 1999 & 2002, along with 'The Patent Rules', 2003.
The Patents Act, (1970). as amended by The Patents (Amendment) Act, (2005) along with The Patents Rules, (2003). As amended by The Patent (Amendment) Rules, (2006).
Tripathi, K. K., (2007). Biotechnology and IPR regime: In: *The Context of India and Developing Countries*, Veena, (ed.), *Biotech. Patent Law*, Hyderabad: ICFAI, pp. 187.
World Intellectual Property Organization, www.wipo.int/wipolex/en/details.jsp?id=7620 (accessed Aug 10, 2016).

CHAPTER 8

GEOGRAPHICAL INDICATIONS IN INDIAN AGRICULTURE

R. C. CHAUDHARY

Chairman, Participatory Rural Development Foundation (PRDF), Gorakhpur (U.P.)–273014, India, E-mail: ram.chaudhary@gmail.com

CONTENTS

8.1 DEFINITION

Geographical indication (GI) of goods, as the name implies, is an indication in the form of a name or a sign for goods that have a specific geographical origin and possess qualities or a reputation that are due to the place of origin. In order to function as a GI, a sign must identify a product as originating in a given place. In addition, the qualities, characteristics, or reputation of the product should be essentially due to the place of origin. Because the qualities depend on the geographical place of production, there is a clear link between the product and its original place of production.

A GI right enables those who have the right to use the indication to prevent its use by a third party whose product does not conform to the applicable standards. For example, in the jurisdictions in which the "Darjeeling Tea" GI is protected (Dattawadkar and Mohan, 2012), producers of Darjeeling tea can exclude use of the term "Darjeeling" for tea not grown in their tea gardens or not produced according to the standards set out in the code of practice for the GI.

However, a protected GI does not enable the holder to prevent someone from making a product using the same techniques as those set out in the standards for that indication. Protection for a GI is usually obtained by acquiring a right over the sign that constitutes the indication.

8.2 GEOGRAPHICAL INDICATION IN AGRICULTURE

Seed or planting material is basic to all agricultural production. Seed costs minimum in the total cost of crop production, but it has the maximum impact. Having reaped the benefit through the seeds of green revolution varieties, farmers were quick to realize the importance of good seeds of new and better varieties of crops. For such superior seeds, farmers were all the more willing to pay a higher price. Seed companies and technology developers saw this as an opportunity to convert plant varieties and important plant genes as profit-making products. Global strategists, pesticide manufacturers, and seed companies joined hands to consolidate capital and technology to dominate the market. In various countries, the need to conserve biodiversity; farm level variation; giving credit to farmers for their traditional crop varieties, folk varieties, and farmers varieties; access to benefit sharing; extending consumer assurance by way of GIs; appellation of origin and traditional knowledge, etc. were attempted to be protected. Global commodity trade is now dominated by several such new issues, which in India are now understood and applied. Another aspect of GI in agriculture is related to the plant-based products or by-products. Plant-based products could be raw material for production or its processing or the preparation. After the GI became effective on September 15, 2003, Darjeeling Tea (Dattawadkar et al., 2012) became the first GI-tagged product in 2004 in India. After that landmark (Gopalakrishnan et al., 2007), many GI-labeled agricultural products have been added (Table 8.1) in India.

Parts of an article or apparatus are, in general, classified with the actual article or apparatus, except where such parts constitute articles included in other classes.

8.2.1 PROTECTION BY GI: LEGAL SIDE

The Indian Parliament enacted in 1999 "The Geographical Indications (GI) of Goods (Regulation and Protection) Act" (Gopalakrishnan et al., 2007) for registration and better protection in relation to goods (www.ipindia.nic.in/girindia/journal/journal_1 to 70.pdf). This Act came into effect on September 15, 2003. Under Section 1(e), GI in relation to goods implies an indication that identifies goods such as agricultural goods, natural goods, or manufactured goods as originating or manufactured in the territory of a country or a region or a locality in that territory, where a given quality reputation or other characteristic of such good is essentially attributed to its geographical origin and in case where such goods are manufactured goods, one of the activities of either the production or of processing or preparation of the goods concerned takes place in such territory, region, or locality as the case may be. The focus of the Act is on quality reputation or other characteristic of such goods, which is essentially attributed to their geographical origin. In doing so, the geographical domain can be a territory of a country or a region or locality in that territory. The quality of the product is attributed essentially to its geographical origin, and if it is goods, then either the raw material production or processing or the preparation shall take place in such territory. The Registrar of the GI shall con-

TABLE 8.1 Classification of Goods– Name of the Product in Each Class

Class	Products/Goods
Class 1	Chemical used in industry, science, photography, agriculture, horticulture and forestry; unprocessed artificial resins, unprocessed plastics; manures; fire extinguishing compositions; tempering and soldering preparations; chemical substances for preserving foodstuffs; tanning substances; adhesive used in industry
Class 2	Paints, varnishes, lacquers; preservatives against rust and against deterioration of wood; colorants; mordents; raw natural resins; metals in foil and powder form for painters; decorators; printers and artists
Class 3	Bleaching preparations and other substances for laundry use; cleaning; polishing; scouring and abrasive preparations; soaps; perfumery, essential oils, cosmetics, hair lotions, dentifrices
Class 4	Industrial oils and greases; lubricants; dust absorbing, wetting and binding compositions; fuels(including motor spirit) and illuminants; candles, wicks
Class 5	Pharmaceutical, veterinary and sanitary preparations; dietetic substances adapted for medical use, food for babies; plasters, materials for dressings; materials for stopping teeth, dental wax; disinfectants; preparation for destroying vermin; fungicides, herbicides
Class 6	Common metals and their alloys; metal building materials; transportable buildings of metal; materials of metal for railway tracks; non-electric cables and wires of common metal; ironmongery, small items of metal hardware; pipes and tubes of metal; safes; goods of common metal not included in other classes; ores
Class 7	Machines and machine tools; motors and engines (except for land vehicles); machine coupling and transmission components (except for land vehicles); agricultural implements other than hand-operated; incubators for eggs
Class 8	Hand tools and implements (hand-operated); cutlery; side arms; razors
Class 9	Scientific, nautical, surveying, electric, photographic, cinematographic, optical, weighing, measuring, signaling, checking (supervision), life saving and teaching apparatus and instruments; apparatus for recording, transmission or reproduction of sound or images; magnetic data carriers, recording discs; automatic vending machines and mechanisms for coin-operated apparatus; cash registers, calculating machines, data processing equipment and computers; fire extinguishing apparatus
Class 10	Surgical, medical, dental and veterinary apparatus and instruments, artificial limbs, eyes and teeth; orthopedic articles; suture materials
Class 11	Apparatus for lighting, heating, steam generating, cooking, refrigerating, drying ventilating, water supply and sanitary purposes
Class 12	Vehicles; apparatus for locomotion by land, air or water
Class 13	Firearms; ammunition and projectiles; explosives; fire works

TABLE 8.1 (Continued)

Class	Products/Goods
Class 14	Precious metals and their alloys and goods in precious metals or coated therewith, not included in other classes; jewellery, precious stones; homological and other chronometric instruments
Class 15	Musical instruments
Class 16	Paper, cardboard and goods made from these materials, not included in other classes; printed matter; bookbinding material; photographs; stationery; adhesives for stationery or household purposes; artists' materials; paint brushes; typewriters and office requisites (except furniture); instructional and teaching material (except apparatus); plastic materials for packaging (not included in other classes); playing cards; printers' type; printing blocks
Class 17	Rubber, gutta percha, gum, asbestos, mica and goods made from these materials and not included in other classes; plastics in extruded form for use in manufacture; packing,= stopping and insulating materials; flexible pipes, not of metal
Class 18	Leather and imitations of leather, and goods made of these materials and not included in other classes; animal skins, hides, trunks and travelling bags; umbrellas, parasols and walking sticks; whips, harness and saddlery
Class 19	Building materials, (non-metallic), non-metallic rigid pipes for building; asphalt, pitch and bitumen; non-metallic transportable buildings; monuments, not of metal.
Class 20	Furniture, mirrors, picture frames; goods(not included in other classes) of wood, cork, reed, cane, wicker, horn, bone, ivory, whalebone, shell, amber, mother- of-pearl, meerschaum and substitutes for all these materials, or of plastics
Class 21	Household or kitchen utensils and containers(not of precious metal or coated therewith); combs and sponges; brushes(except paints brushes); brush making materials; articles for cleaning purposes; steel wool; unworked or semi-worked glass (except glass used in building); glassware, porcelain and earthenware not included in other classes
Class 22	Ropes, string, nets, tents, awnings, tarpaulins, sails, sacks and bags (not included in other classes) padding and stuffing materials(except of rubber or plastics); raw fibrous textile materials
Class 23	Yarns and threads, for textile use
Class 24	Textiles and textile goods, not included in other classes; bed and table covers.
Class 25	Clothing, footwear, headgear
Class 26	Lace and embroidery, ribbons and braid; buttons, hooks and eyes, pins and needles; artificial flowers
Class 27	Carpets, rugs, mats and matting, linoleum and other materials for covering existing floors; wall hangings (non-textile)

TABLE 8.1 (Continued)

Class	Products/Goods
Class 28	Games and playthings, gymnastic and sporting articles not included in other classes; decorations for Christmas trees
Class 29	Meat, fish, poultry and game; meat extracts; preserved, dried and cooked fruits and vegetables; jellies, jams, fruit sauces; eggs, milk and milk products; edible oils and fats
Class 30	Coffee, tea, cocoa, sugar, rice, tapioca, sago, artificial coffee; flour and preparations made from cereals, bread, pastry and confectionery, ices; honey, treacle; yeast, baking powder; salt, mustard; vinegar, sauces, (condiments); spices; ice
Class 31	Agricultural, horticultural and forestry products and grains not included in other classes; live animals; fresh fruits and vegetables; seeds, natural plants and flowers; foodstuffs for animals, malt
Class 32	Beers, mineral and aerated waters, and other non-alcoholic drinks; fruit drinks and fruit juices; syrups and other preparations for making beverages
Class 33	Alcoholic beverages(except beers)
Class 34	Tobacco, smokers' articles, matches

strue the GI in the Registry (Gopalakrishnan et al., 2007; Jena and Grote, 2012; WTO 2004; WTO 2004).

There are three main ways to protect a GI (WIPO 2004; WTO 2004):

i. using *sui generis* systems (i.e., special regimes of protection);
ii. using collective or certification marks; and
iii. using business practices, including administrative product approval schemes.

These approaches involve differences with respect to important questions, such as the conditions for protection or the scope of protection. On the other hand, two of the modes of protection — namely *sui generis* systems and collective or certification mark systems — share some common features, such as the fact that they set up rights for collective use by those who comply with defined standards.

Broadly speaking, GIs are protected in different countries and regional systems through a wide variety of approaches and often using a combination of two or more of the approaches outlined above. These approaches have been developed in accordance with different legal traditions and within a framework of individual historical and economic conditions. In many *sui generis* legislations, registrations for GI are not subject to a specific period of validity (Belleti and Marescotti, 2008; Taubman, 2001; Thiedig and Sylvander, 2000). This means that the protection for a registered GI will remain valid unless the registration is cancelled. GIs registered as collective and certification marks are generally protected for renewable 10-year periods. The right to use a protected GI belongs to producers in the geographical area defined, who comply with the specific conditions of production for the product.

8.2.2 TRIPS REQUIREMENTS AND GI

Trade-Related Aspects of Intellectual Property Rights (TRIPS) prescribes minimum standards for the protection of GI. Additional protection on wines and spirits were granted under Article 23 of the TRIPS Agreement in the Uruguay Round of WTO negotiations. In the Doha Round, many Member Nations desired extending a similar level of protection to some of their important goods as well. The TRIPS contains two protections standards for GI, and Article 22(2) requires countries to provide a legal means to prevent the use of GI that suggest that the goods originate in a geographic area other than the true place of origin. Article 23(3) requires that countries should keep in place a legal means to invalidate the registration of trademarks, which contain or consist of a GI with respect to goods not originating in the territory indicated. These provisions are applicable only if the use of the GI is such that it leads to misleading the public as to the true place of origin of the product. Article 24 states that a GI does not have to be protected if it has not been protected or ceases to be protected in the country of origin or when it is generic term for a product.

8.2.3 TRADEMARK AND GI

GIs identify a good as originating from a particular place. By contrast, a trademark identifies a good or service as originating from a particular company. A trademark (TM) often consists of a fanciful or arbitrary sign. In contrast, the name used as a GI is usually predetermined by the name of a geographical area. Finally, a TM can be assigned or licensed to anyone, anywhere in the world, because it is linked to a specific company and not to a particular place. In contrast, a GI may be used by any persons in the area of origin, who produces the good according to specified standards, but because of its link with the place of origin, a GI cannot be assigned or licensed to someone outside that place or not belonging to the group of authorized producers.

While TM indicates that the product is affiliated with the manufacturer, the GI indicates to the consumer the high quality and reputation of the produce coming from a defined geographical area. The GI can be used by all producers in the area along with their TM. But as a rule, TM that contains a GI cannot be protected, if the use of the TM misleads the public about the true origin of the product. The development of GI is a time-tested process, and to carve an aurora about the product, it takes decades if not centuries. GI creates a positive impression of the product quality, the environmental virtue, and human skill of the area. The premium price it fetches occurs in a gentle manner over a protracted period of time and by varies assessment procedures. Only if the GI can create a positive mind frame on the client over the product will the GI be considered to have some virtue. Thus, while extending the use of GI for food products, care should be taken to ensure that the GI strictly complies with all these requirements. Extending the GI for products that is yet to establish a reputation and consumer credibility will dilute the whole purpose of having market dominance and may discredit the entire exercise.

Like all intellectual property rights (IPR), the rights to GI are enforced by the application of national legislation, typically in a court of law. The right to take action could rest with a competent authority, the public prosecutor, or to any interested party, whether a natural person or a legal entity, whether public or private. The sanctions provided for in national legislation could be civil (injunctions restraining or prohibiting unlawful acts, actions for damages, etc.), criminal, or administrative.

8.2.4 SUPERIORITY OF GI

It is important to be able to distinguish between brand names containing a geographical term and a GI. The reason why there is an increased rush for GI is that the GI protects the consumer and also safeguards the interest of the producers. The GI is perceived as both origin and quality indicator because of which the consumer willingly pays a premium price and that leads to the growth of the regional economy. This is evident by the fact that the European Union alone has granted so far more than 5,000 different GIs.

The GIs of goods Act 1999 is intrinsically integrated with Section 3 of the Trademarks Act 1999 (see Section 2(2) of the GI Act 1999). The rational of protecting the GI is similar to that of the intellectual property (IP) production. The TRIPS agreement says "to be eligible for a GI, a good must possess a quality, reputation or other characteristics attributable to its geographic origin." However, there are fundamental differences between TM and GI. TM identifies a manufacturer, implies certain amount of human creativity, and is usable only by one agency or entity. On the contrary, the GI is complex in definition and perception. It denotes the source of origin, where product quality or specialty that the consumer prefers is governed by the specific physical or biological environment. There is no originality or invention or discovery involved, and the GI may depend on traditional knowledge (TK) for that product development or on the talent of the craftsman. The GI can also be used by all those who produce that product in that given area and is not restrictive.

8.2.5 HUMAN IMMIGRANT AND GI

The post World War II period witnessed large-scale migration and settlement of people from old world to the new world countries. These migrants carried with them their ethnic craft and plants to their new-found lands. They even named the new territory provinces, cities, streets, rivers, and mountain after the ones in their "original homeland." With several subsequent minor modifications, many foodstuffs and farm products were marketed in the name of their "original homeland." This situation creates enormous confusion in the market place between original and new settlement products. There is a running global debate on this confusion of GI, and with emotions being high, the issue has become very complicated.

8.2.6 DANGERS OF TOO GENERIC GI

A zone is an area of land without any particular qualifying attribute, but an agro-climatic zone is decided based on similarity on soil, climate, weather, and other edaphic factors. A region is a single tract of land comprising independently owned farmlands, e.g., Northwest India. A region is said to be discrete between adjoining regions with measurable homogeneity. The subregion ensures a substantial level of homogeneity in the attributes of the produce covered under GI. Therefore, there is likely to be minor variation in the product, if the GI area is larger. For example, Basmati rice if granted GI may cover the rice-growing tracts of Northwest India and Pakistan, while there are minor but acceptable levels of variations between Basmati from Amritsar, Karnal/Kurukshetra, and Dehradun for the reason that this rice-growing zone is quite larger and enjoys some variation in climate. The current Basmati definition accommodates certain defined number of varieties, and if the scope of the definition is further enlarged for the purpose of clubbing several of the new rice genotypes that may have Basmati-like characteristics or better grain, then such an action may even defeat the very purpose of seeking market dominance for this product through GI. Basmati remains a disputed product and has not been given GI due to conflicts from within and outside the country.

A name that has become generic means the name of an agricultural product or foodstuff which, although relates to the place or the region where this product was originally produced or marketed, has become the common name of an agricultural product or a foodstuff. To decide if a given GI has become generic, the following factors can be considered:

- Assess the prevailing situation in the Member State in which the GI name originates and the area of consumption of the produce.
- The situation in other Member States on the above parameter is examined.
- The relevant national or community laws should have adequate provisions to govern reputation.

Understanding the generic GI, cases for "Basmati" and the definitions given in the "Export of Basmati Rice (Quality Control and Inspection) Rules 2003" are important. Adding several other new varieties meeting Basmati Export Standard under the Basmati banner would lead to the Basmati GI becoming generic. These new varieties of very high grain quality, with high productivity per hectare can be given another brand name, and brand equity can be promoted. Trade concerns, consumer trust, and maintenance of product quality are the essence of GI, and that would get eroded if the brand Basmati becomes a generic term. Because cultivation of Basmati involves the livelihood security of millions of farmers, rocking the term "Basmati" periodically, with conflicting objectives is not desirable. A generic definition of GI for basmati and "Claw Back" (CB) option of the European Community are to be kept in mind.

The GI used to describe an agricultural product or foodstuff should cover:

- originating in a specific region, place, or country, and possesses a specific quality reputation or other characteristics attributable to that geographical origin and the production and/or processing of which is done in the defined geographical area.
- any established/traditionally valued direct link must exist between the quality or characteristics of the product and its specific geographic origin.

Very often, the GI material are named and misspelled (to fake it) in a manner that consumers are misled. Homonymous indications are those that are spelled and pronounced alike but mean different as the geographical origin of these products or originate from different countries. Conflicts invariably arise when products of homonymous GI are used and sold in the same market. The problem becomes acute if the homonymous GI products are identical in nature. Honesty is business not being a virtue; clandestine branding of GI is a stark violation of trade rules and procedures and now is legally punishable.

The European Community has taken steps to CB certain GI originating in the European Community, such as the TM PARMA that was registered in country of origin as Mexico. The CB of the GI means confiscating TMs without any compensation and without representation from the TM owner during the negotiations. The GI protection therefore calls for multilateral system for the notification and registration of GIs, and the issue of CB of country-approved GIs on the basis of generic terms or trade needs through discussion. A sound international binding on GI matters is required to ensure that TM owners and users of prior generic terms enforce their legal positions properly.

8.2.7 APPELLATION OF ORIGIN/INDICATION OF SOURCE

"Appellation of Origin" (AO) means that a product originates in a specific geographical region and the characteristic qualities of the product are due to the geographical environment, including natural and human factors. Most of the agricultural produce falls under AO. The Lisbon Agreement defines the AO as the geographical name of a country, region, or locality that serves to designate the product originating therein, the quality and characteristic of which are exclusively or essentially due to the geographical environment, including natural and human factors. "Indication of source" means that a product originates in a specific geographical region. The "Indication of Source" is clarified as "'all goods bearing a false or deceptive indication by which one of the country or place of origin shall be seized on importation into any of the said countries." The Lisbon Agreement is considered to be narrow in its scope on AO than the GI now discussed under TRIPS. It is primarily because the AO is not based on the reputation of a product, which also means that the TK is not a requirement for getting AO accredited.

An indication of source can be defined as an indication referring to a country (or to a place in that country) as being the country or place of origin of a product. In contrast to a GI, an indication of source does not imply the presence of any special quality, reputation, or characteristic of the product essentially attributable to its place of origin. Indications of source only require that the product on which the indication of source is used originate in a certain geographical area. Examples of indications of source are the mention, on a product, of the name of a country, or indications such as "made in ....," "product of ....," etc.

Appellations of origin are a special kind of GI. GIs and appellations of origin require a qualitative link between the product to which they refer and its place of origin. Both inform consumers about a product's geographical origin and a quality or characteristic of the product linked to its place of origin. The basic difference between the two concepts is that the link with the place of origin must be stronger in the case of an appellation of origin. The quality or characteristics of a product protected as an appellation of origin must result exclusively or essentially from its geographical origin. This generally means that the raw materials should be sourced in the place of origin and that the processing of the product should also occur there. In the case of GIs, a single criterion attributable to geographical origin is sufficient – be it a quality or other characteristic of the product – or even just its reputation.

Products identified by a geographical indication are often the result of knowledge carried forward by a community in a particular region from generation to generation. Similarly, some products identified by a GI may embody characteristic elements of the traditional artistic heritage developed in a given region, known as "traditional cultural expressions" (TCEs). This is particularly true for tangible products such as handicrafts made using natural resources and having qualities derived from their geographical origin. GIs do not directly protect the subject matter generally associated with TK or TCEs, which remains in the public domain under conventional IP systems. However, GIs may be used to indirectly contribute to their protection, for instance, by preserving them for future generations. This can be done, for example, through the description of the production standards for a GI product, which may include a description of a traditional process or traditional knowledge.

In the context of geographical indications, generic terms are names which, although they denote the place from where a product originates, have become the term customary for such a product. An example of a GI that has become a generic term is Camembert for cheese. This name can now be used to designate any camembert-type cheese.

The transformation of a geographical indication into a generic term may occur in different countries and at different times. This may lead to situations where a specific indication is considered to constitute a geographical indication in some countries, whereas the same indication may be regarded as a generic term in other countries.

Protection may be requested by a group of producers of the product identified by the geographical indication. The producers may be organized as an entity, such as a cooperative or association, which represents them and ensures that the product fulfils certain requirements which they have agreed upon or adhered to. In some jurisdictions, protection may also be requested by a national competent authority (for example, a local government authority). Protection for a GI is granted by a national (regional) competent authority upon request. In some countries, the function of granting GI protection is carried out by a special body responsible for GI protection. In other countries, the national IP office carries out this function. A directory is available on the WIPO website

A sign must qualify as a GI under the applicable law and not be subject to any obstacles to register as a GI. Generally, an important requirement under the definition, is that the good identified by the GI needs to have a link to the geographical origin. This link may be determined by a given quality, reputation, or other characteristic essentially due to the geographical origin. In many legislations, a single criterion attributable to the geographical origin is sufficient, be it a quality or other characteristic of the product, or only its reputation. A request of protection for a GI may be filed, depending on the applicable law, without assistance from an IP lawyer or specialized agent. However, in many countries, an applicant whose residence or principal place of business is outside the country where the protection is sought must be represented by a lawyer or agent permitted to practice in that country. Information on the permitted lawyers and agents may be obtained directly from the national IP offices. A directory of IP offices is available on the WIPO website.

As the costs for filing for protection vary from country to country, it is best to contact the national (regional) IP office for details on the fee structure. If protection abroad is sought, in addition to the ordinary filing fees, one should take into account the translation costs and the costs of using a local agent. It is worth remembering that in order to protect a GI abroad, there may be a requirement to first protect the GI in the country of origin.

The following are generally excluded from GI protection:

- Signs that do not qualify as GIs under the applicable law. From a legal point of view, potential obstacles to successfully registering a geographical indication (GI) may include the following:
- Conflict with a prior mark.
- Generic characteristic of the term that constitutes the GI.
- The existence of a homonymous GI that would mislead as to the product's true origin.
- The indication's name being that of a plant variety or animal breed.
- The lack of protection of the GI in its country of origin.

If the GI protection is limited to the national level, then the first port-of-call should be the relevant IP office or the national (regional) competent authority in charge of GIs. A directory of IP offices is available on the WIPO website. If, however, one is considering protection in more than one territory, then WIPO's Lisbon System could be an appropriate alternative, amongst others. See the question "Can I obtain geographical indication protection that is valid in multiple countries?" for more information and to learn about other alternatives. There is no comprehensive way to search all GIs registered throughout the world. One can, however, contact the relevant national IP office, which may or may not offer a searchable database of GIs registered in their territory. A directory of IP offices is available on the WIPO website. One can also consult WIPO's Lisbon Express database to search GIs registered under the Lisbon System.

One can use the WIPO Lex search engine to browse the IP laws of WIPO, WTO, and UN members. One needs to select the country(ies) one is interested in and choose "geographical indications" as a subject matter filter. In addition, information on GIs may be provided by national or regional IP offices. A directory is available on the WIPO website.

Consumers are paying more and more attention to the geographical origin of products, and many people care about specific characteristics present in the products they buy. In some cases, the "place of origin" suggests to consumers that the product will have a particular quality or characteristic that they may value. GIs therefore function as product differentiators on the market, by enabling consumers to distinguish between products with geographical origin-based characteristics and others without those characteristics. GIs can thus be a key element in developing collective brands for quality-bound-to-origin products. Consult the WIPO Lex database to browse relevant national legislation.

Protecting a GI enables those who have the right to use the indication to take measures against others who use it without permission and benefit from its reputation ("free-riders"). A GI's reputation is a valuable, collective, and intangible asset. If not protected, it could be used without restriction, and its value diminished and eventually lost (AIACA 2011; Das, 2006; Sahai and Barpujari, 2007). Protecting a GI is also a way to prevent registration of the indication as a trademark by a third party and to limit the risk of the indication becoming a generic term. In general, GIs, backed up by solid business management, can bring with them (Datta, 2009; Sahai and Barpujari, 2007):

- competitive advantage;
- more added value to a product;
- increased export opportunities;
- a strengthened brand.

Homonymous geographical indications (GI) are those that are spelled or pronounced alike, but which identify products originating in different places, usually in different countries. In principle, these indications should coexist, but such coexistence may be subject to certain conditions. For example, it may be required that they be used only together with additional information as to the origin of the product in order to prevent consumers from being misled. A GI may be refused protection if, due to the existence of another homonymous indication, its use would be considered potentially misleading to consumers with regard to the product's true origin.

8.2.8 RELATIONSHIP BETWEEN FARMER'S VARIETIES (FV) AND GI

The PPV & FR Act 2001 (www.plantauthority.gov.in/pdf/application%status.pdf) provides certain rights to farmers, such as to save, use, sow, resow, exchange, share, or sell his farm produce including that of the registered variety. Farmers cannot multiply the seeds of the notified variety on their own or market seeds of registered variety as branded seed with packing, label, etc. and such violation may invite infringement action. The Act recognized farmers as plant breeders and therefore has extended the benefit of entitlement for developing commercial varieties though unaided calls for advanced scientific knowledge, access to diverse germplasm and meticulous experimentation to access the commercial potential of the material. Farmers who do develop new varieties of plants like any other plant breeder can apply their material for the conduct of distinctiveness, uniformity, and stability (DUS) testing and registration. This de-centralization of variety development is one benefit that would spin-off from the PPV & FR Act.

The PPV & FR Act 2001 provides breeders certain ownership claim of their variety when the variety fulfill the criteria of Distinctness, Uniformity and Stability. In many cases, uniformity invariably provides a window for the assessment of stability. In open pollinated crops, the uniformity depends on the nature of the inbred line. That apart, it also depends on the plant breeding methodology followed (top cross, two ways cross, etc.). If genetic male sterile systems (GMS) are used in hybrid development, then the level of uniformity may pose a limitation. The private seed companies tend to focus their attention on the endowed areas where farming is efficient, diverse, and productivity levels are high. Crops grown under marginal, suppressive soils or under arid conditions may not get the same type of attention from private breeders. But gradually, over an extended period of time, the benefit reaches out to all farmers.

In the last 100 years, there has been a drive for improved agriculture and that has replaced FV in several crops with new varieties developed by the plant breeders. Yet, FV is still dominant in pulses, vegetables, melons, etc. The GI for agricultural goods like Basmati rice, coffee, tea, wine, etc. revolve around consumer preferences for the palate feeling, aroma, and physical appearance that enhance the appetite. An ideal mixture of all these attributes raises the value of the product due to reasons of consumer preference. India has enacted the GI Act, and a number of agricultural and handicraft products have been given the GI. India should examine the GI for its agricultural produce like Basmati rice, Alphonso mango, etc., seriously to give it a comprehensive protection of the plant material as FV under the PPV & FR Act 2001 and at the same time give GI protection for produce such as rice, mango fruit and fruit products, etc. Such a double coverage will enable IP protection of the plant material and market advantage to the quality produce through GI.

8.3 THE TRACEABILITY ISSUE

The traceability of the raw material that yields the GI produce is important and the detail of the growers and their track record details are a matter of detailed documentation. The GIs are essentially collective marks and are put to use for the collective benefit of the producers in the GI region. Apart from genotype, the cultivation practices and seasonality of various consignments should be within the area range, and the quality of the produce must remain comparable if GI is to be sustained as a trade advantage. This calls for proper survey of the growing area, identifying the farms, documenting their cultivation details, giving them their unique number that can be traced, indicating it in the container of the graded and packed produce, etc. The cost involved in this exercise is to be met by the growers themselves or their organizations. This added expenditure should match the market benefit that farmers will get out of this exercise. The consumer will bear the burden of cost in many of these cases, and he should see that the value provided to his food source traceability and its dependability is acceptable to him. Very often, these requirements are imposed on the produce originating from a developing country by the West, insisting on it as part of the quality assurance drive. But the hidden agenda could be to use this as a non-tariff barrier to discourage imports. To comply with the traceability demand, developing countries have to invest in a high technology and thus would incur an overhead expenditure to sustain their agriculture exports. Hence, it can also lead to multinationals coming in a big way with capital and technology, and they may export India's farm produce.

8.4 PROCESS OF REGISTERING FOR GI

8.4.1 ORGANIZATIONAL STRUCTURE

Under the Department of Industrial Policy and Promotion of the Ministry of Commerce and Industry, the office of the Controller General of Patents, Designs and Trademarks (CGPDTM) function. It main office is located in Mumbai. The head office of the Patent Office is located in Kolkata, and its branch offices are located in Chennai, New Delhi, and Mumbai. The Trademarks registry is located at Mumbai with branches at Kolkata, Ahmedabad, Chennai, Kolkata, and New Delhi. The Design office is located at Kolkata. The offices of the Patent Information System and National Institute of Intellectual Property Management are located at Nagpur. In order to protect the Geographical Indications (Registration and Protection) Act 1999, a Geographical Indications Registry has been established in Chennai under the CPDTM. The Intellectual Property of Office of India based at Chennai handles all the matters related to GI application and operations. Detailed information on GI could be downloaded from the website: www.ipindia.nic.in. While applying for GI, one has to select a particular class (Table 8.1) to which the intended product belongs.

8.4.2 GEOGRAPHICAL INDICATION APPLICATION

The application can be completed online but must be printed for signature and submission. The following information is required:

a) Name of the applicant
b) Address
c) Type of goods
d) Specifications
e) Name of the GI
f) Description of goods
g) Geographical area of production
h) Proof of origin
i) Method of production
j) Uniqueness
k) Inspection body

After completing the application, it should be submitted to: "Geographical Indications Registry, Intellectual Property Office Building, Industrial Estate, G.S.T Road, Guindy, Chennai–600 032, Ph: 044-22502091-93 & 98, Fax: 044-22502090, E-mail: gir-ipo@nic.in, Website: ipindia.gov.in"

The steps and process that follows the application is outlined in Figure 8.1. The validity of GI Registration is for the period of 10 years, which can be revalidated following the same process. Any infringement and unlawful use of GI is punishable under law. In Uttar Pradesh, only few individuals/organizations have come forward for GI registration. Of the 63 GI registered in agriculture, only 3 namely, Allahabadi Surkha guava, Mango Malihabadi Dussehari, and Kalanamak rice, are registered under GI during 2014 (Gopalakrishnan et al., 2007). This is unacceptable situation and government agencies, NGOs, and individuals must take due efforts to improve this situation.

8.5 STORY OF GEOGRAPHICAL INDICATION FOR KALANAMAK RICE IN INDIA

Kalanamak is the famous, prestigious, and heritage rice of eastern Uttar Pradesh, India. An improved variety named Kalanamak KN3 was already

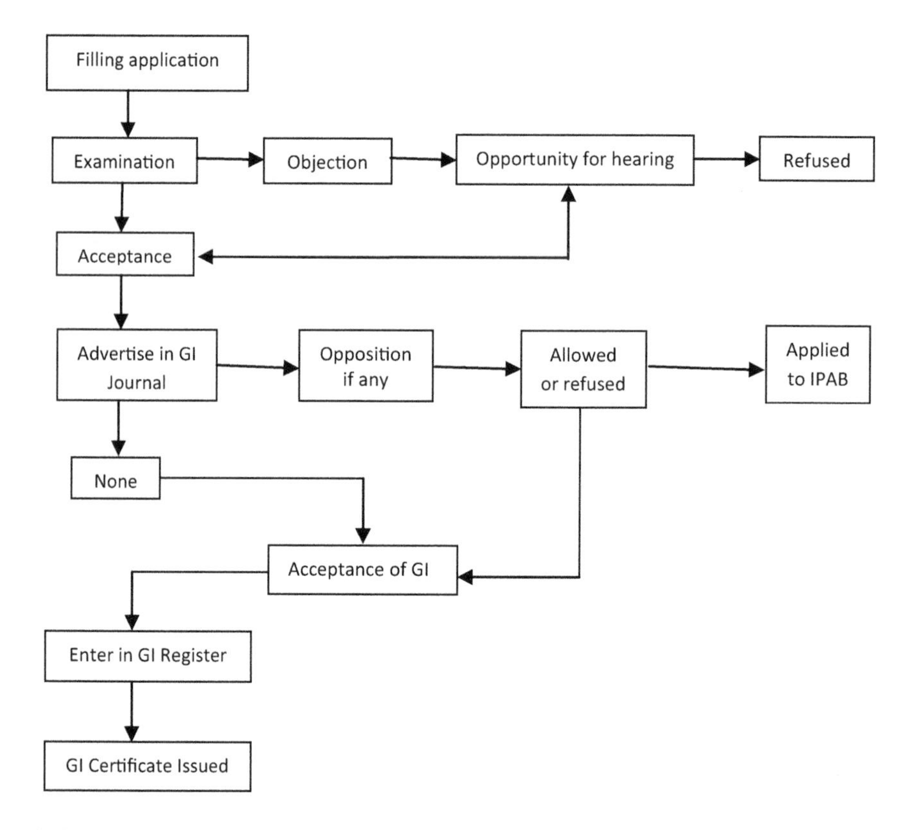

FIGURE 8.1 Steps and processes involved in the registration of geographic indications.

released and notified (Notification of Government of India No. 3 # SO2137 (E) dated August 31, 2013). Kalanamak was also protected under the PPV & FRA Act 2001 (www.plantauthority.gov.in/pdf/application%status.pdf 1117=REG/2009/138) by Participatory Rural Development Foundation (PRDF), Gorakhpur. An NGO based in Siddharth Nagar applied to get GI on Kalanamak. The application was "advertised" on the website (www.ipindia.nic.in) following the procedure that within 3 months, any one could protest or advice on the contrary. PRDF based in Gorakhpur cooperated and pointed out several flaws in the proposal. Description of the Kalanamak variety was incorrect and morpho-agronomic characteristics were completely wrong. The other major flaw was the indicated area for GI; it was merely 5 villages chosen haphazardly from around Naugarh township of Siddharth Nagar district alone. The villages were also not contiguous. This would have been a disaster for Kalanamak rice (Figures 8.2 and 8.3), disaster for the community, and would have triggered civic strife. But all of these was avoided by the timely intervention of PRDF Gorakhpur. GI was granted to Kalanamak rice on September 8, 2013 and published in the 2013–2014 issue of GI News. Now, GI for Kalanamak covers agro-climatic Zone 6 of Uttar Pradesh covering 11 districts, namely Bahraich, Balrampur, Basti, Gonda, Gorakhpur, Deoria, Kushinagar, Mahrajganj, Sant Kabir Nagar, Siddharth Nagar, and Sravasti, located between Nepal border in the north to Ghaghra river in the south, and Bahraich in the west to Deoria in the east.

With general awareness increasing, there is an increasing trend in GI registration (Table 8.1). Details of GI registration issued for agricultural and horticultural products are given in Table 8.2.

FIGURE 8.2 Kalanamak (improved) rice crop grown in the GI area of Uttar Pradesh, India.

FIGURE 8.3 Grain (paddy) and milled rice of kalanamak.

8.6 GI REGISTRATION IN INDIA

India has been slow to start and is still going slow with regard to GI registration (Tables 8.2 and 8.3). During 2003 to 2015, only 75 GI registrations have been done for agricultural goods (Table 8.2) in India, although the country is center of origin of so many plant and animal species. Rich biological diversity abounds India, and the country has history of more than 10,000 years of agriculture. Still, very few GIs have been done. Of 36 states and union territories in India, only '11 have opened their account for GI registration (Tables 8.3–8.5). The trend of GI registration has been slow (Table 8.3) like the trend in popularization of "slow food" (Sople, 2014). There appears no reason other than general lack of awareness about GI even among academicians and institutions (Rangnekar, 2009; Nanda, 2013; Dwivedi and Bhattacharya, 2012; www.foodquality-origin.org/eng/index.html). Individuals do not see immediate economic gain, though it will benefit the country, community, and individuals in the long run (Das, 2009; AIACA, 2011; www.aptdpc.com/home/shows_newsitem/60). Some applications are pending (Table 8.4) as the process of facilitation has also been slow. Still, there is no particular reason why so few applications are

TABLE 8.2 Geographical Indications Registered for Various Agricultural and Horticultural Commodities in India During 2004 to 2015

		Crop/Product	State
Period: 2004 – 2005			
1	Darjeeling Tea (word & logo)	Tea	West Bengal
Period: 2005 – 2006			
2	Kangra Tea	Tea	Himachal Pradesh
3	Coorg Orange	Orange	Karnataka
Period: 2006 – 2007			
4	Mysore Betel leaf	Betel leaf	Karnataka
5	Nanjanagud banana	Banana	Karnataka
Period: 2007 – 2008			
6	Mysore Jasmine	Flower	Karnataka
7	Udupi Jasmine	Flower	Karnataka
8	Hadagali Jasmine	Flower	Karnataka
9	Navara rice	Rice	Kerala
10	Palakkadan Matta rice	Rice	Kerala
Period: 2008 – 2009			
11	Malabar Pepper	Black Pepper	Kerala
12	Allahabad Surkha	Guava	Uttar Pradesh
13	Monsooned Malabar Arabica Coffee	Coffee	Karnataka
14	Monsooned Malabar Robusta Coffee	Coffee	Karnataka
15	Spices – Alleppey Green Cardamom	Cardamom	Kerala
16	Coorg Green Cardamom	Cardamom	Karnataka
17	Eathomozhy Tall Coconut	Coconut	Tamil Nadu
18	Pokkali rice	Rice	Kerala
19	Laxman Bhog Mango	Mango	West Bengal
20	Khirsapati (Himsagar) Mango	Mango	West Bengal
21	Fazli Mango	Mango	West Bengal
22	Naga Mircha	Chillies	Nagaland
23	Nilgiri (Orthodox) Logo	Tea	Tamil Nadu
24	Assam (Orthodox) Logo	Tea	Assam
25	Virupakshi Hill Banana	Banana	Tamil Nadu
26	Sirumalai Hill banana	Banana	Tamil Nadu
27	Mango Malihabadi Dusseheri	Mango	Uttar Pradesh

TABLE 8.2 (Continued)

		Crop/Product	State
Period: 2009 – 2010			
28	Vazhakulam Pineapple	Pineapple	Kerala
29	Devanahalli Pomello	Citrus	Karnataka
30	Appemidi Mango	Mango	Karnataka
31	Kamalapur Red Banana	Mango	Karnataka
32	Bikaneri Bhujia	Product	Rajasthan
Period: 2010 – 2011			
33	Guntur Sannam Chili	Chillies	Andhra Pradesh
34	Mahabaleshwar Strawberry	Strawberry	Maharashtra
35	Central Travancore Jaggery	Jaggery sugar	Kerala
36	Wayanad Jeerakasala Rice	Rice	Kerala
37	Wayanad Gandhakasala Rice	Rice	Kerala
38	Nashik Grapes	Grape	Maharashtra
39	Byadgi Chilli	Chillies	Karnataka
40	Gir Kesar Mango	Mango	Gujarat
41	Bhalia Wheat	Wheat	Gujarat
Period: 2011 – 2012			
42	Udupi Mattu Gulla Brinjal	Brinjal	Karnataka
43	Ganjam Kewda Rooh	Kewda Flower	Odisha
44	Ganjam Kewda Flower	Kewda Flower	Odisha
45	Madurai Malli	Jasmine Flower	Tamil Nadu
Period: 2012 – 2013			
46	Bangalore Blue Grapes	Grape	Karnataka
Period: 2013 – 2014			
47	Kalanamak Rice	Rice	Uttar Pradesh
48	Kaipad Rice	Rice	Kerala
49	Kolhapur Jaggery	Jaggery sugar	Maharashtra
50	Nagpur Orange	Orange	Maharashtra
51	Thanjavur Rice Maalai	Rice	Tamil Nadu
52	Vangurla Cashew	Cashew	Maharashtra
53	Sangli Resins	Grape	Maharashtra
54	Lasalgaon Onion	Onion	Maharashtra

TABLE 8.2 (Continued)

		Crop/Product	State
55	Beed Custard Apple	Custard Apple	Maharashtra
56	Jalna Sweet Orange	Orange	Maharashtra
57	Sangli Turmeric	Turmeric	Maharashtra
58	Ratnagiri Alphonso Mango	Mango	Maharashtra
59	Jalgaon Banana	Banana	Maharashtra
60	Marathwada Kesar Mango	Mango	Maharashtra
61	Purandar Fig	Fig	Maharashtra
62	Jalgaon Bharit Brinjal	Brinjal	Maharashtra
63	Solapur Pomegranate	Pomegranate	Maharashtra
Period: 2014 – 2015			
64	Bangalore Rose Onion	Onion	Karnataka
65	Naga Tree Tomato	Tomato	Nagaland
66	Arunachal Orange	Orange	Arunachal
67	Sikkim Large Cardamom	Cardamom	Sikkim
68	Mizo Chilli	Chilli	Mizoram
69	Assam Karbi Anglong Ginger	Ginger	Assam
70	Tripura Queen Apple	Apple	Tripura
71	Chengalikodan Nendran Banana	Banana	Kerala
72	Tezpur Litchi	Assam	Assam
73	Khasi Mandarin	Mandarin	Meghalaya
74	Kachai Lemon	Lemon	Manipur
75	Memong Narang	Orange	Meghalaya

Source: GI Journal No. 1 to 70 of years 2004–2015.

filed annually. Among the states, the maximum number of GI (17) has been registered from Karnataka, followed by Maharashtra (Table 8.5). None of the 7 union territories have opened their account. Limited awareness has been generated by the concerned government agencies (AIACA, 2011; www.aptdpc.com/home/shows_newsitem/60). GI has not only economic (Ganguli, 2009) and social benefits (Das, 2006; Jain, 2009), but it also protects the national wealth from being unduly exploited by others. Moreover, it protects TK and germplasm (Rangnekar, 2004; Nair and Kumar 2005; Sahai and Barpujari 2005, www.origin-food.org/2005/

TABLE 8.3 Trend in Number of Registration Issued for Geographical Indication in a Decade (2004–2015) for Agricultural and Horticultural Products in India

S. N.	Period	Number of Registration
1	April 2004 – March 2005	1
2	April 2006 – March 2007	2
3	April 2006 – March 2007	2
4	April 2007 – March 2008	5
5	April 2008 – March 2009	17
6	April 2009 – March 2010	5
7	April 2010 – March 2011	10
8	April 2011 – March 2012	4
9	April 2012 – March 2013	1
10	April 2013 – March 2014	17
11	April 2014 – March 2015	12

upload/SIN-WP1-report-131006.pdf) of unique quality. Thus, GI is valuable (Vandecandelayere et al., 2011; www.iprsonline.org/resources/Geographical_Indications.htm) and imperative for any individual, community, and country.

8.7 GEOGRAPHICAL INDICATIONS FOR DEVELOPING COUNTRIES

GI is a powerful tool to protect the ownership right on natural resources, natural product, and byproducts based on plants and animals. It started with the developing countries like UK, France, and others. Now, these countries are even registering their products like Scotch whisky and Champagne in developing countries like India under International Registration. They want to protect their products being labeled elsewhere. For the developing countries, GI is a boon as it has limited costs and simpler procedure compared to the other forms of IPR. The entire community in the GI area can benefit. There is no hidden cost or hidden danger to it. After 10 years, GI can be renewed indefinitely. Thus, developing countries must proceed with GI before someone else can stake their claim on

TABLE 8.4 Applications Pending Before GI Registry as on March 2015

Application Number	GI for	Status	Date of filing	GI area
14	Basmati Rice	Refused	19.08.2004	Punjab/Haryana
139	Alphonso Mango	Examination	22.09.2008	Maharashtra
145	Basmati	Opposed	26.11.2008	India*
206	Rataul Mango	Examination	05.04.2010	U.P.
231	Erode Turmeric	Examination	04.01.2011	Tamil Nadu
245	Naga Cucumber	Examination	25.11.2011	Nagaland
370	Jhabua Kadaknath black chicken	Examination	08.02.2015	M.P.
379	Devgad Alphonso Mango	Examination	16.03.2012	Gujarat
401	Mahoba Pan	Examination	05.02.2013	U.P.
406	Salem Mango	Examination	03.05.2013	Tamil Nadu
407	Hosur Rose	Examination	03.05.2013	Tamil Nadu
439	Joha Rice of Assam	Pre Exam.	10.09.2013	Assam
464	Sirsi Siddapur Yellapur Arecanut	Pre Exam.	10.12.2013	Karnataka
470	Ajara Ghansal Rice	Pre Exam.	26.03.2014	Maharashtra
471	Waigaon Turmeric	Pre Exam.	26.03.2014	Maharashtra
472	Mangalwedha Maldandi Jowar	Pre Exam.	26.03.2014	Maharashtra
473	Bhiwapur Chilli	Pre Exam.	26.03.2014	Maharashtra
476	Waghaya Ghevada	Pre Exam.	26.03.2014	Maharashtra
477	Navapura Desi Tur	Pre Exam.	26.03.2014	Maharashtra
478	Marlashi Ambemohar Rice	Pre Exam.	26.03.2014	Maharashtra
484	Thanjavur Rice Maalai	Pre Exam.	23.05.2014	Tamil Nadu
489	Vengural Cashew	Pre Exam.	22.07.2014	Maharashtra
490	Sangli Raisins	Pre Exam.	22.04.2014	Maharashtra
491	Lasalgaon Onion	Pre Exam.	22.07.2014	Maharashtra
493	Ghalvad Chikoo	Pre Exam.	11.08.2014	Maharashtra
494	Beed Custard Apple	Pre Exam.	26.08.2014	Maharashtra
495	Jalna Sweet Orange	Pre Exam.	26.08.2014	Maharashtra

TABLE 8.4 (Continued)

Application Number	GI for	Status	Date of filing	GI area
496	Sangli Turmeric	Pre Exam.	26.08.2014	Maharashtra
497	Ratnagiri Alphonso Mango	Pre Exam.	26.08.2014	Maharashtra
498	Jalgain Banana	Pre Exam.	26.08.2014	Maharashtra
499	Marathwada Kesar Mango	Pre Exam.	30.09.2014	Maharashtra
500	Purandar Fig	Pre Exam.	30.09.2014	Maharashtra
501	Jalgaon Bharit Brinjal	Pre Exam.	30.09.2014	Maharashtra
520	Uttarakhand Ka Tejpat	Pre Exam.	27.01.2015	Uttarakhand
525	Bardhman's Sitalbhog	Pre Exam.	13.03.2015	W.B.
526	Bardhman's Mihidana	Pre Exam.	13.03.2015	W.B.

* Delhi, Haryana, Himachal, Jammu & Kashmir, Punjab, Uttarakhand, U. P. *www.ipinidca.nic.in*

it. Because GIs are embedded in a territory, they can be effective tools for promoting local knowledge and local development. They can also be protected in many countries by International Registration using the Madrid System. Mexican Tequila, Thai Silk, Nepal Himalayan Tea, and Darjeeling Tea are many such examples of GI multicountry registered products from developing countries.

TABLE 8.5 State-wise Distribution of GI Done in India During 2003 to 2015

S. N.	Name of the State	No. of GI done
1	Andhra Pradesh	1
2	Arunachal Pradesh	1
3	Assam	3
4	Gujarat	2
5	Himachal Pradesh	1
6	Karnataka	17
7	Kerala	11
8	Maharashtra	16
9	Manipur	1
10	Meghalaya	2
11	Mizoram	1
12	Nagaland	2
13	Odisha	1
14	Rajasthan	1
15	Sikkim	1
16	Tamil Nadu	6
17	Tripura	1
18	Uttar Pradesh	3
19	West Bengal	4
Total		**75**

KEYWORDS

- **appellation of origin**
- **Darjeeling tea**
- **IPR**
- **protection of landraces**
- **WTO**

REFERENCES

Addor, F., & Grazioli, A., (2002). Geographical indications beyond wine and spirits. *J. World Intellect. Property*, *5*(6).

All India Artisans and Craftworkers Association (AIACA), (2011). Geographical indications of India: socioeconomic and development issues, Policy Brief, New Delhi, AIACA.

Andhra Pradesh shows the way forward for Geographical Indications, www.aptdpc.com/home/shows_newsitem/60.

Anon., (2009). Geographical indications its evolving contours. SVKM's NMIMS University, Mumbai, India, pp. 87. www.iips.ac.in.

Belleti, G., & Marescotti, A., (2008). Geographical indications strategies and policy recommendations. SINER-GI EU funded project, *Final Report*, Toulouse (F), website: www.origin-food.org/).

Das, K., (2006). Protection of India's geographical indications, an overview of the Indian legislation and the TRIPS scenario. *Ind. Jour. International Law, 46*(1), 39–72.

Das, K., (2008). *Geographical Indications: UNCTAD's Initiative in India*, presentation at UNDP RCC, UNDP Cambodia and Economic Institute of Cambodia, Phnom Penh. http://hdru.aprc.undp. org/ext/regional_workshop_2008/pdf/Das_ s3.

Das, K., (2009). *Socio-Economic Implications of Protecting Geographical Indications in India*. New Delhi: Centre for WTO Studies. http://wtocentre.iift.ac.in/ Papers/GI_Paper_CWS_August%2009_Revised. pdf, accessed on January 5, 2012.

Datta, T. K., (2009). *Tea Darjeeling*, India, FAO Case study, pp. 113–188.

Dattawadkar, N., & Mohan, H., (2012). The status of geographical indications in India: A short review. *Intellectual Property Division*, Stellarix Consultancy Pvt. Ltd. India, pp. 1–4.

Dwivedi, K., S., & Bhattacharjya, S., (2012). Restore glory of the Banarasi sari, *The Hindu Business Line*. http://www.thehindubusinessline.com/ opinion/restore-glory-of-the-banarasi-sari/ article4226412.ece.

FAO Quality linked to geographical origin: www.foodquality-origin.org/eng/index.html.

Ganguli, P., (2009). GI, its evolving controls, *WTC Research Studies Report,* pp. 4.

GI Journal: www.ipindia.nic.in/girindia/journal/journal_1 to 70.pdf.

Gopalakrishnan, N. S., Nair, P. S., & Babu, A. K., (2007). Exploring the relationship between geographical indications and traditional knowledge: an analysis of the legal tools for the protection of geographical indications in Asia, *Working Paper*, Geneva, International Centre for Trade and Sustainable Development (ICTSD), pp. 1–65.

Jain, S., (2009). Effect of the extension of Geographical Indications: A South Asian perspective, *Asia Pacific Development Jour.*, *15*(2).

Jena, P. R., & Grote, U., (2007). Changing institutions to protect regional heritage, *A Case for Geographical Indications in the Indian Agri. Food Sector*. http:// www.pegnet.ifw-kiel.de/activities/pradyot.pdf, accessed on February 2, 2012.

Nair, L. R., & Kumar, R., (2005). Geographical indications – a search for identity, *Butterworth*, New Delhi, pp. 95.

Nanda, N., & Barpujari, I., (2012). Traditional knowledge and limits to GI, *The Hindu Business Line*.

Nanda, N., (2013). The protection of geographical indication in India: Issues and Challenges. *TERI Briefing Paper,* New Delhi, pp. 12.

Rangnekar, D., (2004). The socio-economics of geographical indications, *Bridges Between Trade and Sustainable Development, 8*(8), 20–21.

Rangnekar, D., (2009). Geographical indications and localization: A case study of Feni United Kingdom: *Centre for the Study of Globalization and Regionalization*, University of Warwick, pp. 1–62.

Sahai, S., & Barpujari, I., (2007). Are geographical indications better suited to protect indigenous knowledge? *A Developing Country Perspective*, New Delhi: Gene Campaign. http://www.genecampaign.org/home_files/Gene_Briefing/Policy%20Brief-2.pdf, accessed on December 22, 2012.

SINER-GI, (2006). WP1 report: Legal and institutional issues related to GIs. www.origin-food.org/2005/upload/SIN-WP1-report-131006.pdf.

Sople, V. V., (2014). *Managing Intellectual Property: The Strategic Imperative.* PHI Learning Private Ltd., Delhi, India. Fourth Edition, pp. 1–107.

Taubman, A., (2001). "The way ahead: Developing international protection for geographical indications: thinking locally, acting Globally." Lecture, *WIPO Symposium on the International Protection of Geographical Indications,* Montevideo, pp. 12. www.wipo.int/edocs/mdocs/geoind/en/wipo_geo_mvd_01/wipo_geo_mvd_01_9.pdf.

Thiedig, F., & Sylvander, B., (2000). Welcome to the club? An economical approach to geographical indications in the European union. (Agricultural Economics) *Agrarwirtschaft, 49, Heft, 12*, pp. 428–437.

UNCTAD: United Nations Conference on Trade and Development, www.iprsonline.org/resources/Geographical_Indications.htm.

Vandecandelaere, E., Arfini, F., Belletti, G., & Marescoti, A., (2010). *Linking People, Places and Products*, 2nd edition, FAO, Rome, Italy, pp. 194.

WIPO, (2004). *Geographical Indications: An Introduction*, publ. no. *489*. WIPO, pp. 44.

WIPO: World Intellectual Property Organization/Appellations of Origin: www.wipo.int/lisbon/en.

WTO, (2004). *Exploring the Linkage between the Domestic Policy Environment and International Trade*, http:// www.wto.org/english/res_e/booksp_e/anrep_e/ world_trade_report04_e.pdf, accessed on January 16, 2013.

CHAPTER 9

PROTECTING NON-BASMATI INDIGENOUS AROMATIC RICE VARIETIES OF WEST BENGAL, INDIA UNDER GEOGRAPHICAL INDICATION: A CRITICAL CONSIDERATION

KOUSHIK ROY,[1] ANIRBAN MUKHERJEE,[2,3] ANIRUDDHA MAITY,[4] KUMARI SHUBHA,[5] and ARINDAM NAG[6]

[1] *RRS (Hill Zone), Uttar Banga Krishi Viswavidyalaya, Kalimpong, Darjeeling, West Bengal–734301, India*

[2] *Social Science Section, ICAR-Vivekananda Parvatiya Krishi Anusandhan Sansthan, Uttarakhand–263601, India, E-mail: anirbanmujkiari@gmail.com*

[3] *Division of Agricultural Extension, ICAR-Indian Agricultural Research Institute, New Delhi–110012, India*

[4] *Division of Seed Technology, ICAR- Indian Grassland and Fodder Research Institute, Jhansi, U.P. – 284003, India*

[5] *Germplasm Evaluation Division, ICAR-National Bureau of Plant Genetic Resources, New Delhi–110012, India*

[6] *Dr. Kalam Agricultural College, Bihar Agricultural University, Kishanganj, Bihar – 855107, India*

CONTENTS

ABSTRACT

Indigenous landraces of aromatic rice in India have been facing serious threat with the introduction of miracle seeds of high-yielding varieties and chemical intensive farming on the backdrop of green revolution. The report says that more than 100 indigenous aromatic rice varieties used to be cultivated in different regions of undivided Bengal. Many more landraces are still under cultivation in some small pockets of the state by personal endeavor of counted farmers but are not documented properly. Surprisingly, at present, only a few indigenous aromatic rice varieties are being cultivated in some scattered pockets of different districts in West Bengal, though in an unorganized way. Erosion of the huge rice genetic biodiversity including aromatic rice poses a serious concern for the farming community, environmentalists, scientists, policymakers, and especially rice-loving consumers. Therefore, considering the importance of these precious heirloom varieties, necessary strategic research and promotion are to be streamlined to protect from erosion. For the proper promotion of such non-basmati landraces (*Gobindabhog, Tulaipanji, Badshabhog, Dudheswar Seetabhog, Radhatilak, Radhunipagal, Kalonunia, Kalojira,*

Tulsimukul, etc.), geographical indication (GI) can be an important instrument. Acceptability of such rice varieties by the retailers and consumers in international and domestic markets is high; hence, promotion of farmers' producer organization in this area would be a much-needed effort. Here, one of the important issues to be addressed is benefit sharing with traditional knowledge holder and improvement of their socio-economic condition. Therefore, GI can be a platform for product and market development addressing the socio-economic issue of the beholder. In this chapter, an attempt has been made to highlight the present issue, status, and scope of non-basmati aromatic varieties of West Bengal to be registered under GI.

9.1 INTRODUCTION

Genetic diversity of landraces and its wild relatives provides the foundation of evolution and its differentiation into various cultivars to adapt in different environments for any crops. But assimilation in the culture of tribes and folks is the special and rare features of cereals like cultivated rice (*Oryza sativa*). In South Asia, more than 1,00,000 folk landraces of the *indica* rice were found to be distributed in remote villages (Richharia and Govindasamy, 1990). Such rich treasures vanishing with time due to changes in gastronomic preferences and culinary practices in different food cultures. Furthermore, this astounding genetic diversity began to decline during the 1970s with the so-called green revolution when the then modern high-yielding varieties (HYVs) were introduced with grain yield enhancement as the primary objective (Shiva, 1991; Dwivedi, 1997; Deb, 2005). About 5000 rice varieties were shipped to the International Rice Research Institute (IRRI) from north east part of India in 1965, but none of this "Assam Collection" survives in Assam and the surrounding states (Jackson, 1994). In West Bengal, over 5500 varieties were recorded to have existed until the 1970s, of which about only 3500 varieties were shipped to the IRRI (Deb, 2005). Not only India but also other south Asian countries have witnessed the same situation. In Bangladesh, about 6600 such varieties were replaced by modern HYVs (Thrupp, 2000). Similarly, the number of local varieties has drastically declined in Taiwan, China, Japan, and Southeast Asian countries, owing to a shift to the monoculture of modern varieties since the 1970s (Gao, 2003). During the end of the last century, we have lost about 75% of crop genetic diversity of the world, as farmers have discarded their heirloom varieties for genetically uniform HYVs (Gliessman, 2007).

Aromatic rice assumes immense importance and is preferred by consumers all over the world due to its flavor and palatability; thus, it fetches higher market price than good quality nonaromatic rice. Because of the quality and auspiciousness, the aromatic or scented rice have occupied a prime position in Indian society (Ahuja et al., 1995). Among the scented rice, the basmati type is accepted as the best scented, longest, and slenderest rice in the world. Basmati rice is also highly regarded throughout Asia and is becoming popular in Europe (Berner and Hoff, 1986) and the USA and in nontraditional rice-growing countries like Australia (Blakeney, 1992). Worldwide, the popular aromatic rice varieties are mostly long-grained, although a majority of indigenous aromatic rice varieties (IARVs) in India are small and medium-grained (Singh et al., 2000). Previously, a large number of such aromatic rice varieties were collectively called Basmati (bas=aroma), but over the years, the designation of basmati has changed, and it is increasingly limited to long slender grain type with moderate to strong aroma (Mahindru, 1995). Accordingly, the small- and medium-grain aromatic rice varieties are being regarded as a separate class of non-basmati aromatic rice.

India is a country with an enormous wealth of these rice varieties and races. But a lot has already been lost as an aftermath of the Green Revolution where the major emphasis was on yield rather than on quality (Singh and Singh, 1998). This is despite the fact that some of the non-basmati scented rice is much superior to basmati types with respect to traits like aroma, kernel elongation after cooking, fluffiness, taste, etc. Nevertheless, due to the special attachment of the farmers with basmati rice, a large number of them are still in existence.

Almost every state in India has its own varieties of aromatic rice, and there is specificity of an area as per cultivation and aroma formation. These areas are identified from hundreds of years of experience of farmers (Nene, 1998). The state of West Bengal had a wealthy source of IARVs with small and medium grain of excellent grain quality including aroma. Prior to Green Revolution, farmers used to nurture diverse rice varieties in their fields. The report says that more than 100 indigenous aromatic rice varieties were cultivated in different regions of undivided Bengal (still, much more landraces are cultivated in some small pockets), but at present, only 12–15 indigenous aromatic rice varieties are being cultivated in an unorganized manner in some scattered pockets of different districts in West Bengal. Erosion of the huge rice genetic biodiversity including aromatic rice poses a serious concern for the farming community, environmentalists, scientists, policymakers, and rice consumers.

Geographical indication (GI) has traditionally been considered to be intellectual property (IP). Article 1(2) of the Paris Convention for the Protection of Industrial Property of 1883 (Paris Convention) refers to "indications of source" and the "appellations of origin" to be considered as objects of industrial property. GIs are *distinctive signs* used to differentiate rival goods of the same category. They are collectively owned by the geographical region which they refer with a strong inherent *origin*-base. The reference to geographical origin – most regularly for agricultural products – combined with the use of traditional methods of extraction and further processing, presents an interesting marketing potential in terms of product branding. Therefore, considering the importance of these precious heirloom varieties, in this chapter, we have tried to show not only an option to protect such variety from erosion, but also to promote livelihood through the legislative tool GI.

9.2 CLASSIFICATION OF RICE VARIETIES: BASMATI AND NON-BASMATI

In the India, scented rice varieties are categorized as basmati and non-basmati. The basmati rice is characterized by long slender grains with kernel length of 6 mm and above, length to breadth ratio (L/B ratio) of 3 and above, and high kernel elongation after cooking. The grains are pointed at both ends with gradual tapering at the end opposite to the germination end and have uniform breadth between the tapering (Mahindru, 1995). The non-basmati aromatic rice varieties also have one or more of the basmati traits, although it does not possess all traits. Especially, they have small and medium kernel length and kernel elongation after cooking is not as much as basmati. As a highly profitable product, basmati received more importance than non-basmati type. Today, basmati shares a maximum portion of export under the aromatic rice category, and the amount is too high. Though basmati rice constitutes a tiny portion of the total rice produced in India, but by volume, the share of Basmati is around 6% (as of FY2016), and by value, it accounts for 60% (as of FY2016) of India's total rice exports. The export value of basmati rice has increased from Rs. 10,890 crores in FY2010 to Rs. 22,718 crores in FY2016 at a compounded annual growth rate (CAGR) of 13%. The commercial benefit has driven stakeholders to focus on Basmati, and in the process, non-basmati has gradually lost its space in the competition (Table 9.1).

In India, several ICAR institution and agricultural university research systems have developed renowned basmati varieties and shall continue to develop in future also. The following basmati rice varieties are notified and released for commercial interest in India (Table 9.2).

A plenty of aromatic rice varieties has been described in several ancient literature and cultural manuscripts. Several varieties were not cultivated and yet may have vanished from the civilization pathway. Still, there are more than 1000 non-basmati indigenous aromatic rice varieties in India, and some of these are of worth to cultivate commercially. Based on the grains size, the rice varieties are classified as large, medium, and small (Table 9.3). It is prominent that the majority of the non-basmati indigenous aromatic rice varieties are small grained, whereas the basmati varieties are generally long grained. As the commercial consumers prefer long-grained varieties more than medium and small-grained varieties, the commercial importance of the latter varieties is reducing day after day. Aroma and texture are considered as important factor for commercial valuation. But,

TABLE 9.1 Morphological Classification of Basmati and Non-Basmati Scented Rice

Characteristics	Non-basmati scented	Basmati type
Length to Breadth ratio	3.2-3.3	3.5-4.2
Kernel length (in mm)	5.3-5.7	6.4-7.6
Length of cooked grain to Breadth of cooked grain ratio	2.9-3.7	4.9-5.6
Breadth of cooked grain to Breadth of uncooked grain ratio	1.3-1.6	1.26-1.33
Length of cooked grain to Length of uncooked grain ratio	1.4-1.6	1.7-1.8

TABLE 9.2 Notified Varieties of Basmati Rice in India

Sl. No.	Basmati Varieties	Name of the Breeding Institutions	No. & Date of Notification
1	Basmati 217	Punjab Agricultural University, Panjab	NA
2	Basmati 370	Rice farm, Kalashah Kaku (Pakistan)	361(E)-30.06.1973 786-02.02.1976
3	Type 3 (Dehraduni Basmati)	Rice Research Station, Nagina, Uttar Pradesh	13-19.12.1978
4	Taraori Basmati(HBC 19/ Karnal Local)	CCSHAU, Haryana	1(E)- 01.01.1996
5	Ranbir Basmati	Rice Research Station, R. S Pura, Jammu and Kashmir	1(E)-01.01.1996
6	Basmati 386	Punjab Agricultural University, Panjab	647(E)-09.09.1997
7	Punjab Basmati 1(Bauni Basmati)	Punjab Agricultural University, Panjab	596(E)-13.08.1984
8	Pusa Basmati 1	ICAR-Indian Agricultural Research Institute, Delhi	915(E)-06.11.1989
9	Kasturi	ICAR-Directorate of Rice Research, Rajendra Nagar, Hyderabad, Telangana	915(E)-06.11.1989
10	Haryana Basmati 1	CCSHAU, Haryana	793(E)-22.11.1991
11	Mahi Sugandha	Rice Research Station, Banswara, Rajasthan	408(E)-04.05.1995
12	Pusa Basmati 1121 After amendment	ICAR-Indian Agricultural Research Institute, Delhi	1566(E)-05.11.2005 2547(E)-29.10.2008
13	Improved Pusa Basmati 1(Pusa 1460)	ICAR-Indian Agricultural Research Institute, Delhi	1178(E)-20.07.2007
14	Vallabh Basmati 22	SVBU AT, Modipuram, Uttar Pradesh	2187(E)-27.08.2009
15	Pusa Basmati 6 (Pusa 1401)	ICAR-Indian Agricultural Research Institute, Delhi	733(E)-01.04.2010
16	Punjab Basmati 2	Punjab Agricultural University, Panjab	1078(E)-26.07.2012

TABLE 9.2 (Continued)

Sl. No.	Basmati Varieties	Name of the Breeding Institutions	No. & Date of Notification
17	Basmati CSR 30 After amendment	ICAR-Central Soil Salinity Research Institute, Karnal, Haryana	1134(E)-25.11.2001 2126(E)-10.09.2012
18.	Pusa Basmati 1509	ICAR-Indian Agricultural Research Institute, Delhi	2817(E)-19.09.2013
19.	Malviya Basmati Dhan	Banaras Hindu University, Varanasi, Uttar Pradesh	2817(E)-19.09.2013
20.	Vallabh Basmati	SVBPUAT, Modipuram, Uttar Pradesh	2817(E)-19.09.2013
21.	Basmati 564	Sher-e-Kashmir University of Agricultural Sciences and Technology , Jammu and Kashmir	268(E)-28.01.2015
22.	Vallabh Basmati 23	SVBPU AT, Modipuram, Uttar Pradesh	268(E)-28.01.2015
23.	Vallabh Basmati 24	SVBPU AT, Modipuram, Uttar Pradesh	268(E)-28.01.2015

these are influenced by environmental effects like duration of sun shine, temperature, RH and rainfall etc. For example, the rice variety *Gopalbhog* grown in Bihar and West Bengal differ in the aroma. Similarly the rice variety Tulaipanji flourishes its aroma with full potential if only cultivated in Dinajpur districts of West Bengal. This highlights the importance of geography and GI.

9.3 STATUS OF SOME DOMINANT AREAS OF NATIVE AROMATIC RICE IN INDIA

A survey of the Seola-Majra belt of Dehradun district of Uttarakhand state, once famous for producing the finest quality Basmati rice in the world, indicated that near 80% of the prime basmati growing area in this belt is now utilized for housing (Singh et al., 1997). Lal basmati of Tapovan, a village above Lakshman Jhoola in Rishikesh district, vanished due to loss

TABLE 9.3 Non-Basmati Indigenous Aromatic Rice Varieties of Different States of India

States	Long grains	Medium grains	Small grains
Andhra Pradesh		Jeeragasambha	
Bihar	Baikani	Bhilhi Basmati, Amod, Kalanamak, Kesar, Sonachur	
		Champaran Basmati(Bhuri), Gopalbhog, Champaran Basmati(Lal), Abdul, Bahami,Champaran Basmati (Kali)	
Assam			Krishna Joha, Kunkuni Joha, Manikimadhuri Joha, Boga Tulsi, Bogi Joha, Bokul Joha, Borjoha, Borsal, Cheniguti, Chufon, Bengoli joha, Bhaboli Joha, Kalijeera, Kamini Joha, Kataribhog, Khorika Joha, Kola Joha, Bhuguri, Boga Joha, Bogaminiki Madhuri, Goalporia Joha-1, 2, Kon Joha-1,2, Ramphal Joha, Ranga Joha Govindbhog, Joha Bora, Koli Joha
Haryana	Basmati, Pakisthani Basmati Basmati 370, Khalsa 7, Tararoi		
Himachal Pradesh	Chimbal Basmati, Mushkan, Seond Basmati Baldhar basmati, Madhumalati	Panarsa Local Achhu, Begmi	
Kerala			Gandhakasala, Jeerkasala

TABLE 9.3 (Continued)

States	Long grains	Medium grains	Small grains
Madhya Pradesh	Laloo	Vishnu Parag, Chatri, Madhuri	Vishnubhog, Badshabhog, Chinore, Dubraj, Kalu, Mooch, Tulsi Majari
Karnataka		Kagasali	
Maharashtra		sakoli-7, Kagasali, Prabhavati	Chinore, Ambemohor
Punjab	Pakisthani Basmati, Basmati 385, Basmati 370		
Rajasthan	Basmati 370, Basmati (Local)		
West Bengal		Katanbhog, Kanakchur	Kataribhog, Radhunipagol, Sitabhog, Badshabhog, Chinisakkar, Kalonunia, Tulaipanji, Danaguri, Tulsibhog Gandheswari
Uttar Pradesh	Dubraj, Basmati 370, Dehraduni Basmati, Type 3, Hansraj, Kalasukhdas, Lalmati, Tapovan Basmati, T-9, Nagina 12, Duniapat(T9), Ramjinwain (T1) Safeda, Vishnu Parag	Tilak Chandan, Kesar, Karmuhi, Keasr, Prasam, Kalanamak, Vishnuparag, Sonachur,	Kanak Jeeri, Laungchoor, Moongphali, Rambhog, Ramjawain, Adamchini, Bindi, Chhoti Chinnawar, Dhania, Bengal Juhi, Thakurbhog, Yuvaraj, Bhantaphool, Jeerabattis, Badshapasand, Bhanta Phool, Sakkarchini, Tinsukhia

of purity and poor fertility of the soil. The Tarai belt adjoining to Nepal border and comprising Chanpatia, Ramnagar, Narkatia, and Jhumka area of west Champaran in Bihar were once considered as the bowel of aromatic rice. Several varieties of aromatic rice like Champaran basmati (Lal, Bhuri, and Kali), Kanakjeera, Kamod, Baharni, Dewta bhog, Kesar, Tulsi Pasand, and Badshabhog were once popular in this area. In the last 20 years, rice was the major Kharif crop of this district. Presently, more than half of the area has been replaced by sugarcane (Singh et al., 2000).

Dehradun basmati survived as it was exported. Farmers were getting assured return from the export, and farmers' emotional attachment was the main reason behind its sustenance.

Interested farmers of some of the major rice-growing states, viz., West Bengal, Orissa, Bihar, Chhattisgarh, and Madhya Pradesh, are believed to have conserved some fantastic landraces with unique characteristics of commercial and breeding value. Many of these, with the unique identity, are actually grown with a vernacular name, which often indicates their unique feature (Chakrabarty et al., 2012). The state of West Bengal (WB) had a rich source of small and medium grain IARVs with excellent grain quality including aroma. Before Green Revolution, farmers used to cultivate different rice varieties in their fields.

Over the last decade, the genetic instability of the acquired traits in modern rice varieties and the associated environmental deterioration have caused an unceasing decline in yields and quality of food grains in several Asian countries. This has led to great anxiety among the policymakers and administrators who aim to meet the food requirements of the expanding population. India is currently experiencing a rice/food crisis due to the erosion of its biodiversity and increase in monocropping in agriculture.

India's ban on export of non-basmati rice has been a boon to competitors like Thailand whose shipments have increased by 1.7 million tons in 2008, according to US Department of Agriculture (USDA). "Thailand, Vietnam, Pakistan, Burma and Brazil collectively expanded exports by 1.7 million tons in 2008, much of which went to historically Indian markets in Africa," the USDA (2008). Besides Thailand, the world's biggest rice exporter are now Vietnam and Pakistan; consequently, India has been knocked off to the fourth position. India had banned non-basmati rice exports in April 2008 to contain rising inflation. It, however, permitted shipping abroad of about 2 million tons via diplomatic channels (PTI, 2009).

9.4 STATUS OF WEST BENGAL IN CONSERVING NON-BASMATI RICE LANDRACES

Rice is the staple food in WB and in India. The lateritic region of WB is one the largest agro-climatic regions of this state, comprising six major districts of the state. From time immemorial, this region is one of the largest reservoirs of agro-biodiversity. In 1975, more than 5000 farmers variety of rice was reported from WB, and a large fraction of the varieties was found in this region. But unfortunately, presently, this number has radically dropped down and is limited to less than around four to five hundred varieties only. Likewise, only 200–250 farmers' variety of rice is currently documented from the lateritic zone. The Amarkanan rural socio-environmental welfare society, WB, is engaged in conserving 150 farmers' varieties among them in their repository. These farmers varieties consist of various important agromorphic characteristics and are important economically as well as from the ecological point of view. From the extensive survey, it was observed that the number of farmers variety of rice was declining very fast, and instant attention will be needed; otherwise, we may lose this rich biodiversity, which will be the main reason for a future food crisis. Crop improvement and production of disease-free variety solely depend on upon the selected genes possibly present in the gene pool of farmer's variety of rice (Sinha, 2014). It is believed that many of these varieties, being native to the concerned areas, assumed greater significance in the context of climate change, which has renewed the necessity in the evaluation, conservation, and seed multiplication of such varieties (Chakrabarty et al., 2012).

Well-known Indian rice researchers, Richharia and Govindasamy (1990), in the book "Rices of India" provided Vedic evidence, and the present-day literature show that the country had been gifted with above 2 lakhs (200,000) rice varieties, a rich biodiversity that no other country on the earth possesses.

India abounds with scores of indigenous aromatic short grain cultivars and land races, grown in pockets in different states. Most of the scented rice is highly area-specific; hence, each Indian state has its own special scented rice. Some aromatic or scented traditional or folk varieties of WB are Badshabhog, Radhunipagol, Kataribhog, Kalonunia, Sitabhog, Mahishadan, Tulaipanji, Gandheswari, Tulsibhog, and Gobindabhog. The traditional varieties like Gobindabhog, Badshabhog, or Sitabhog are used to prepare "payesh" a traditional food item in WB. Many states have a collection of native popular scented varieties that are known for their cultivation and adaptation and particularly Uttar Pradesh, Orissa, Bihar, Madhya Pradesh, Assam, and WB have a wealth of genetic divergence in short grain aromatic types.

The major reason for the loss of thousands of local rice varieties is their steady replacement with the high-yielding varieties (HYVs) introduced in the 1960s. Farmers were impressed by the initial high yields of these so-called "miracle" seeds and ignored the associated expenditure on external inputs, subsequent loss of non-grain biomass, loss of desirable traits (like tolerance to disease pests, floods, drought, etc.), and the extensive environmental degradation, including soil and water.

In households in West Midnapore district, Banshkathi and Sitasail folk varieties are cooked to entertain special guests. Muri or puffed rice is a popular snack, especially in rural Bengal. Kelas, Moogi, Dahar Nagra, Nalpai, and Moul folk rice varieties are generally used for making "muri." During winter season, the Bengalis (people of WB) are fond of a special aromatic sweet item known as "Joynagar moa" made of scented "khoi" or popped rice. This "khoi" is made from the folk rice variety Kanakchur, which is mostly grown in the Joynagar area of South 24 Parganas, West Bengal. The variety Kanakchur is conserved by the people because there is a huge demand for the "Joynagar moa."

Various cultural rituals have contributed to the conservation of several folk varieties. Husked and/or dehusked rice is always present in even the simplest Hindu "puja" as one of the offerings and which is also often directly associated with human prosperity and fertility; hence, the custom of throwing rice on newlyweds is widely practiced. Annaprashana is the first rice feeding ceremony of a newborn baby in Bengali culture. The rice for this ceremony is usually a scented traditional variety like Gobindabhog, Badshabhog or Mohonbhog. Gobindabhog rice is an aromatic rice variety that grows in WB. Outside WB, Hyderabad city, Telangana, is one of the primary markets for Gobindabhog rice. In Andhra Pradesh and Telangana, the demand for this variety is for preparing Biryani. Since 2009, there has been a 30% increase in the selling price of Gobindabhog in Murshidabad because of its demand. Currently, this variety fetches a price of Rupees 2,800–3,000 per bag of 60 kg. Therefore, many farmers in WB are opting to cultivate aromatic rice. (Figure 9.1) A total of 80% of the rice produced is consumed locally, while 20% is exported (Pramanik, 2013).

In each and every district of WB, non-basmati aromatic rice varieties are still cultivated in pockets of 50-31000 hectares. There is no specific published data on the extent of area cultivated under folk rice varieties

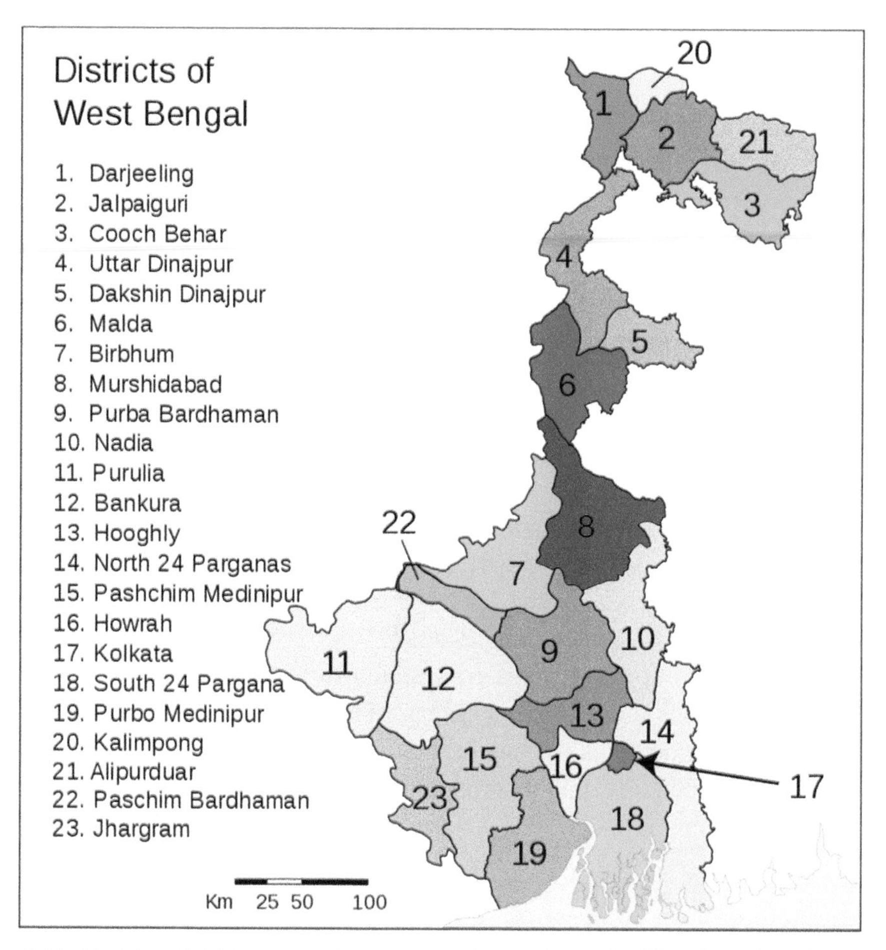

FIGURE 9.1 Districts conserving aromatic rice varieties in West Bengal (Adapted from https://fi.wikipedia.org/wiki/L%C3%A4nsi-Bengalin_piirikunnat#/media/File:West BengalDistricts_numbered.svg https://creativecommons.org/licenses/by-sa/3.0/).

in WB. The data (Table 9.4) were collected from various sources. Dudheswar, Kalonunia, Gobindabhog, and Tulaipanji are some of the promising varieties of Bengal, which are not only famous in India but are also exported in large quantity. Besides the abovementioned area of near about 73000 ha under folk rice varieties, other varieties are still cultivated sporadically.

The agro-morphological and biochemical traits of scented rice landraces of WB are comparable to the other non-basmati aromatic rice

TABLE 9.4 Details on Some Traditional Rice Varieties Being Cultivated in Different Parts of West Bengal

Landraces	Areas (in ha) under cultivation (approx.)	Growing Districts	Utility
Kalturay, Sanuaddey, Addey, Japaka	150	Darjeeling hills	Scented rice used for sweet milk-based dessert 'payesh', pulao, biryani
Sadanunia, Kalonunia,	20,000	Coochbehar, Jalpaiguri	Scented rice used for Making offering to God *(Bhog)* and 'payesh'
Tulaipanji, Kataribhog	6,000	Uttar and Dakshin Dinajpur	'payesh', pulao, biryani and pitha (home made cakes).
Gobindabhog, Badshabhog, sitabhog	31,000	South and North 24 Pgs., Hoogly, Bankura, Burdwan, Birbhum	*'payesh', pulao, biryani* and *pitha*
Marichshal, Khejurchari	150	North and South 24 Parganas	Daily cooking, Panta (water soaked fermented rice) Muri (rice bubble)
Gheus, Kaminibhog, Khoijhuri,	30	Sundarban areas	'payesh', pulao
Kabirajsal, Chamarmani	50	Midnapur, Nadia	*Bhog* and 'payesh'
Jamainaru, Moulo	35	Howrah	*Bhog* and 'payesh'
Radhatilak	50	Hoogly, North 24 Pgs, Nadia, Burdwan, Bankura, and East Midnapur	*Bhog* and 'payesh'
Dudheswar	15,000	South and North 24 Parganas	*Bhog* and 'payesh'
Kalojira	50	Jalpaiguri, Nadia and East Midnapur	*Bhog* and 'payesh'
Tulsimukul	50	Bankura, Birbhum and Purulia	*Bhog* and 'payesh'
Radhunipagal	200	Birbhum, Bankura and west Midnapur	'payesh', pulao, biryani

varieties in India. The growing period varies from 130–150 days and is mostly during the onset of monsoon, i.e., mid-June to July, with grain length of 5–9 mm, kernel L/B ratio of 1.7 to 2.5, and a diverse kernel color (Table 9.5).

All these varieties are still grown in farmers' field because of the traditional cultural practices and rituals followed in WB. In different festival and rituals, the scented rice is used for sweet milk-based dessert "payesh" for offering to god. *Pitha* (homemade cakes), "payesh," pulao, and biryani are common dishes in different festivals. These traditional practices have still kept these varieties alive. But with the urbanization and modernization, people are gradually abandoning their cultures, and with this situation, the majority of traditional non-basmati aromatic rice is facing the risk of extinction like others.

9.5 THE CONTEXT OF FARMERS' ORGANIZATION AS AN OPTION

With the changing markets, the subsistence farming systems are rapidly transformed to input intensive commercial entrepreneurship. Most of the developing countries are now experiencing the paradigm shift from subsistence agriculture to commercial agri-business (Mukherjee et al., 2012a). In this context, the role of technology transfer (extension) has become utmost important. The public agricultural extension service has conventionally been providing the important linkage between agricultural research and farming communities, especially in the transfer of technology and support of agricultural and rural development programs (FAO, 2007). There is increasing demand for rapid input, service, and technology delivery among the farmers, although fulfillment of these demands solely by public extension system is limited. The reason are wide farmer extension workers ratio, i.e., 2879:1 (Mukherjee and Maity, 2015), administrative and bureaucratic workload and financial limitations etc. All these made the public extension services more supply driven rather than demand driven (Sulaiman et al., 2005).

Private extension has been providing some specialized services like real-time information of weather and market and related advisory services based on weather and price forecasting techniques, for exam-

TABLE 9.5 Agro-Morphological and Biochemical Traits of Some Scented Rice Landraces of West Bengal

Variety/landraces	Kernel L/B ratio	Kernel color	Time of sowing	Growing days	Grain length (mm)
Badshabhog	2	Gold and gold furrows on straw background	Mid June	140	5–6
Tulaipanji	2.5	Gold and gold furrows on straw background	Mid June	140	8
Gobindabhog	2	Gold and gold furrows on straw background	Mid June- July	145	5–6
Kaminibhog	1.8	Brown spots on straw	June-July	155	5–6
Kalonunia	2.5	Black	Mid June	150	6–7
Kalojira	1.7	Black	Mid June	150	5–5.5
Lalbadshabhog	2.0	Brown (tawny)	Mid June	130	5.5–6
NC-324	2.5	Purple spots/furrows on straw	Mid June	145	6–7
Kataribhog	2.5	Gold and gold furrows on straw background	Mid June	145	6–7
NC-365	2.5	Black	Mid June	150	8–9
Sitabhog	2.0	Gold and gold furrows on straw background	Mid June	150	6–7
Radhunipagal	2	Purple spots/furrows on straw	Mid June	150	5–6
Radhatilak	2.0	Purple spots/furrows on straw	Mid June	145	6–6.5
Tulsimukul	1.9	Black	Mid June	150	6

ple, soil and water testing followed by customized fertilizer and crop nutrition services, effective market linkages, price-based timely input delivery, etc. (Mukherjee et al., 2012b). The accountability of private extension service has improved with committed and quality services. Although like others, the private extension system is also facing several constraints in functional and organizational structure such as huge workload of ground level workers and their job security issues, high business and profit orientation of companies, high price of products and focus on high-value crops, etc. (Mukherjee et al., 2012c). These companies are focusing more on progressive farmers by avoiding whole village approach which is sometime dangerous in context of equity. It certainly has improved the focus on agribusiness but at the same time has increased social disparity (Singh, 2008). Therefore, many professionals in academic and social research institutes have doubted the intentions of some of the private extension systems. There is rising concern that the farmers' organizations can act as a potential driving force for agricultural and rural development. Farmers' organizations are working as "engines" of development and that can uphold the pennon of development even ahead of the local level, while offering benefits to the rest of society (Blokland, 2007). Through the community and organizational effort, a new model can be established through farmers' organization where registration under GI can promote the rice cultivars and organization can start growing and marketing rice.

9.6 EXISTING GIS IN RICE

A GI is a name or sign used on the products that correspond to a particular geographical location or origin (e.g., area, region, or country). India, as a member of the World Trade Organization (WTO), in 1999 enacted the Geographical Indications of Goods (Registration and Protection) Act 1999, which came into effect from September 15, 2003. Presently, 240 GIs have been registered for different products in India. Of these, 7 are registered for only rice and 6 of them come from Kerala (Table 9.6). The state of WB is far behind despite having such potential varieties to be registered.

TABLE 9.6 Rice Variety Under GI

Sl no	Products	State	Specialty	Unique
1	Navara rice	Kerala	Medicinal properties	Shape, color and size of grain
2	Pokkali rice	Kerala	Organic nature	Symbiotic cultivation with prawn
3	Palakkadan Matta Rice	Kerala	coarseness and health benefits	Brown rice, distinct earthy flavor because of the type of soil in which it is cultivated
4	Wayanad Gandhakasala Rice	Kerala	scented variety, high nutritional value	agro-ecological conditions, the methods of organic cultivation, the traditional genetic make-up of cultivars, and unique processing technologies that have produced the specific aroma and flavor
5	Wayanad Jeerakasala Rice	Kerala	scented variety, high nutritional value	The uniqueness of this rice is mainly attributed to particular climatic conditions prevalent in the area, together with varietal characters and system of rice cultivation, adding to the best expression of aroma and flavor in the product
6	Kalanamak Rice	Uttarakhand	Scented Paddy, salted in taste	Known for its aroma and distinct taste
7	Kaipad Rice	Kerala	salinity-prone natural organic rice	distinct morphological and qualitative traits, and have different salinity tolerance mechanisms, imparting varietal diversity to the unique ecosystem

Under Article 22(1) of the WTO Agreement on Trade-Related Aspects of Intellectual Property Rights (TRIPS) Agreement, GIs have been defined as: "Indications which identify a good as originating in the territory of a member, or a region or a locality in that territory, where a given quality, reputation or characteristic of the good is essentially attributable to its geographic origin." But uniqueness of a product is foremost important for registration as GI. For that conceptual understanding of the process and scientific thoroughness is required. In the case of French wines, the elements of terroir include climate, soil, topography, and associated plants growing in the vicinity, which impart uniqueness and specialty to the product. However, two major components are very much important for Indian agricultural products, viz., geographical link and reputation link (Soam and Hussain, 2011). Product specialty must be the outcome of geographical factors, although product distinctiveness or uniqueness may be the outcome of the combination of geographical and other non-geographical factors.

The basic concept of the GIs lies in integrating the community members for a cause and building confidence among them as the owner of a brand, leading to profit sharing. Profit sharing is very much important for safeguarding the interest of the community. GI is such a tool which not only can safeguard that but can provide a new recognition to the product, which will safeguard the existence of the product. In this regard, the indigenous aromatic rice varieties have tremendous potential to get registered under GI. Social business, scientific rigor and legal technicalities are required to channelize these products into commercially viable ventures through GI portfolio, so that the local producers are encouraged, honored, and benefitted from being protector and proprietors of intellectual property.

9.7 CONCLUSION

Since the green revolution, indigenous landraces of aromatic rice in India have been facing serious threat, and as a result, huge deterioration of rice genetic biodiversity occurred, which further poses a serious concern for the farming community. A very few folk varieties are remaining in the field. All these varieties are still cultivated in farmers' field because of the traditional cultural practices and rituals followed in West Bengal. But how long? These varieties have tremendous potential in the national and global market, although they are still to come under the limelight. Registration under GI can bring them in limelight. To achieve this, social awareness and scientific rigor are the basic needs to be catered first.

KEYWORDS

- **basmati rice**
- **GI**
- **IPR**
- **landraces**
- **protection of local cultivars**

REFERENCES

Ahuja, S. C., Pawar, D. V. S., Ahuja, U., & Gupta, K. R., (1995). *'Basmati Rice –The Scented Perl'*, Directorate of publication, CCS Haryana Agricultural University, Hissar, India, pp. 63.

Berner, D. K., & Hoff, B. J., (1986). Inheritance of scent in American long grain rice. *Crop Science*, *26*, 876–878.

Blakeney, A. B., (1992). Developing rice varieties with different texture and tastes.' *Chem. Australia*, *1*, 475–476.

Blokland, K., & Gouet, C., (2007). Farmers' peer-to-peer support path to economic development. In: Ton, G., Bijman, J., & Oorthuizen, J., (eds.), *Producer Organizations and Market Chains*. Facilitating trajectories of change in developing countries. Wageningen, Wageningen Academic Publishers, 71–88.

Chakrabarty, S. K., Joshi, M. A., Singh, Y., Maity, A., Vashisht, V., & Dadlani, M., (2012). Characterization and evaluation of variability in farmers' varieties of rice from West Bengal. *Ind. J. Gen. Pl. Breed.*, *72*(2), 136–142.

Deb, D., (2005). *Seeds of Tradition, Seeds of Future: Folk Rice Varieties from East India.* Research Foundation for Science Technology & Ecology, New Delhi, pp. 11.

Dwivedi, J. L., (1997). Conserving genetic resources and using diversity in a flood-prone ecosystem in eastern India. In: Sperling, L., & Loevinsohn, M., (eds.), *Using Diversity Enhancing and Maintaining Genetic Resources On-farm*. IDRC, Ottawa. URL: http://www.idrc.ca/books/focus/833/dwivedi.html.

FAO, (2007), Sustainable Agriculture and Rural Development (SARD) *Policy Brief 12*.

Gao, L. Z., (2003). The conservation of Chinese rice biodiversity: genetic erosion, ethno-botany and prospects. *Genetic Resources and Crop Evolution, 50*, 17–32.

Gliessman, S. B., (2007). *Agroecology: The Ecology of Sustainable Food Systems*. CRC Press, Boca Raton/ London.

Jackson, N. E., (1994). Preservation of rice genes. *Nature, 371*, 470.

Mahindru, S. N., (1995). *Manual of Basmati Rice*. Metropolitan Book Co. Pvt. Ltd., New Delhi, India, 307.

Mukherjee, A., & Maity, A., (2015). Public–private partnership for convergence of extension services in Indian agriculture. *Current Science, 109*(9), 1557–1563.

Mukherjee, A., Bahal, R., Burman, R. R., & Dubey, S. K., (2012). Factors contributing farmers' association in Tata Kisan Sansar: a critical analysis. *Indian Res. J. Extension Educ., 12*(2), 81–86.

Mukherjee, A., Bahal, R., Burman, R. R., & Dubey, S. K., (2012a). Conceptual convergence of pluralistic extension at Aligarh district of Uttar Pradesh. *Journal of Community Mobilization and Sustainable Development, 7*(1&2), 85–94.

Mukherjee, A., Bahal, R., Burman, R. R., Dubey, S. K., & Jha, G. K., (2012b). Constraints in privatized agricultural technology delivery system of Tata Kisan Sansar, *J. Glob. Commun., 5*(2), 155–159.

Nene, Y. L., (1998). Basmati rice: a distinct variety (cultivar) of the Indian subcontinent'. *Asian Agri.-History, 2*(3), 175–188.

Paul, A., (2014). Save our rice campaign, *Paddy*, No. *21 & 22*, pp. 5. http://ofai.org/wp-content/uploads/2011/04/Paddy-Save-Our-Rice-Campaign.pdf. (accessed 12 September, 2016).

Pramanik, A., (2013). *Bengal's Aromatic Gobindabhog Rice Attracts more Farmers*. Business Line print edition. http://www.thehindubusinessline.com/economy/agri-business/bengals-aromatic-gobindo-bhog-rice-attracts-more-farmers/article5440868.ece (accessed 10 September, 2016).

PTI, (2009). Rivals gain from India's non-basmati export ban. http://www.business-standard.com/article/markets/rivals-gain-from-india-s-non-basmati-export-ban-109071400007_1.html. (accessed 8 September, 2016).

Richharia, R. H., & Govindasamy, S., (1990). *Rices of India*. Academy of Development Science, Karjat, pp. 18–21.

Roy, K., (2015). Evaluation of agro-technique and storage methods for aromatic rice landraces West Bengal. *PhD Dissertation*, Bidhan Chandra Krishi Viswavidyalaya, pp. 35–92.

Shiva, V., (1991). *The Violence of the Green Revolution*. Third world network, Penang, pp. 5–26.

Singh, A. K., & Narain, S., (2008). Effectiveness of public and private extension system in delivering services. *Indian Res. J. Extension Educ., 8*(2&3), 51–55.

Singh, R. K., & Singh, U. S., (1998). Indigenous scented rices of India: a survival issues. Sustainable agriculture for food, energy and industry. In: Bassam, N. E., Behl, R. K., & Prohnow, B., ed., *Proc. International Conference Held in Braunschweig*, Germany, pp. 676–681.

Singh, R. K., Singh, U. S., & Khus, G. S., (1997). Indegenous aromatic rice of India: Present scenario and needs. *Agricultural Situation in India*, 491–496.

Singh, R. K., Singh, U. S., Khus, G. S., Rohilla, R., Singh, J. P., Singh, G., & Shekhar, K. S., (2000). Small and medium grained aromatic rices of India. *Aromatic Rices. Ed.,* Chapter no 9, pp. 158.

Sinha, A. K., (2014). Threatened traditional rice (*Oryza sativa,* L.) varieties of lateritic region of West Bengal– status, distribution and conservation. *International Journal of Applied Biosciences*, *2*(2), 111–116.

Soam, S. K., & Hussain, M., (2011). Commercialization of indigenous health drinks as geographical indications. *Journal of Intellectual Property Rights*, *16*, 170–175.

Sulaiman, V. R., Hall, A., & Suresh, N., (2005). Effectiveness of private sector extension in India and lessons for the new extension policy agenda. *Agricultural Research & Extension Network, Paper No. 141,* pp. 1–11.

Thrupp, L. A., (2000). Linking agricultural biodiversity and food security: the valuable role of sustainable agriculture. *International Affairs*, *76*(2), 265–281.

CHAPTER 10

AN AGRICULTURAL PERSPECTIVE ON PATENTABILITY OF GENES

RAVI S. SINGH,[1] CHANDAN ROY,[1] TRIBHUWAN KUMAR,[2] CHANDRA S. PRABHAKER,[3] UJJWAL KUMAR,[1] and PRABHASH K. SINGH[1,2]

[1] *Department of Plant Breeding and Genetics, Bihar Agriculture College (BAU), Sabour–813210, India, Tel.: +91-7781896931, E-mail: ravi.molbio@gmail.com*

[2] *Department of Molecular Biology and Genetic Engineering, Bihar Agriculture College (BAU), Sabour–813210, India*

[3] *Department of Entomology, Bihar Agriculture College (BAU), Sabour–813210, India*

CONTENTS

ABSTRACT

The patentability of genes has been a controversial issue that impacts ethics, legality, and scientific research and development globally. Agricultural research and development do not remain untouched, as many private firms like Aventis, Dow Chemical, DuPont, Monsanto, and Syngenta are now holding a large chunk of agricultural patents. The example of gene patenting issues involving Bt Cotton (Monsanto, USA) and Golden rice (developed by Swiss and German scientists) have raised world's ethical and legal concerns. One of the apprehensions is that patent granted for genes will give monopoly to institutions and private companies; as consequence, a huge sum of money will be charged by them as license fee for use by others. The counter argument put forth is that it saves time, funds, and effort of two or more groups working on the same gene. The patent, in essence, is a safeguard against copying, selling, and making profit out of others' innovations involving huge investment of money, time, and human efforts. At the same time, it hinders the progress of scientific research and development by debarring the access to scientific data and research material, which otherwise could be of use for further development. Taken together, pros and cons of patenting genes with agricultural perspective, it should be ethically guided and human-welfare oriented rather than targeted toward making profits.

10.1 INTRODUCTION

The patentability of genes has been a controversial issue that impacts ethics, legality, and scientific research and development globally. Plant Breeders' Rights (PBRs) was the only intellectual property (IP) protection in most developed nations until about 1980 to empower a seed owner for selling seeds of specific seed variety or breeding material.

But recent trend of gene patenting in plants has rendered agricultural research and development very challenging. This is because many private firms like Aventis, Dow Chemical, DuPont, Monsanto, and Syngenta are now holding a large chunk of agricultural patents. The example of gene patenting issues involving Bt Cotton (Monsanto, USA) and Golden rice (developed by Swiss and German scientists) have raised world's ethical and legal concerns. One of the apprehensions is that patent granted for genes will give monopoly to institutions and private companies; as consequence, a huge sum of money will be charged by them as license fee for use by others. The counter argument put forth is that it saves time, funds, and effort of two or more groups working on the same gene.

Trade Related Intellectual Property (TRIPS) agreement signed in 1994 as part of the Uruguay Round of trade negotiations by all nations, including developing nations, committed them to protect plant varieties. In US, patents related to rice surged to more than 600 patents issued annually between the period 1999 and 2000, from below 100 per year in 1995; these comprised patents for specific lines of herbicide-resistant rice (Barton and Berger, 2001). There are broadly claimed US and European patents, covering specific technical procedures used in agricultural genetic engineering of plant varieties by private firms like Agracetus (acquired by Monsanto), which may seek for all transgenic cotton and soybeans. These patents, if valid, could give, control of all transgenic varieties of these crops. The patent, in essence, is a safeguard against copying, selling, and making profit out others' innovations involving huge investment of money, time, and human efforts. At the same time, it hinders the progress of scientific research and development by debarring the access to scientific data and research material, which otherwise could be of use for further development. Taken together, pros and cons of patenting genes with agricultural perspective, it should be ethically guided and human-welfare oriented rather than targeted toward making profit.

10.2 HISTORICAL DEVELOPMENT OF IP LAWS

Patent is an exclusive right given to the inventor for his creation to exclude others to take unlawful benefits out of the creation. Recognizing the inventors is an old age system in the history of inventions. Among the earliest documentation in 600 B.C. Jacob, R. has given IP right on some kind of "newfangled loaf" of bread. Later, Worlds' first legally bound patent system was adopted as Vinetean Act in 1441. This was followed by the British statute of monopoly in 1624. At that time, the inventor was allowed to take the benefits of their invention for 10 years. Austria-Hungary government called for international exhibition in 1873 at Vienna where many international inventors participated in the program. This led to the development of Paris Convention in 1883 for the protection of industrial properties. The word patent originates from Latin word *Patere* means "to lay open." Patent is the oldest form of intellectual property rights (IPR). Patent grants only for those activities that are new, developed through inventive steps, and have industrial applicability. Anything that is mere discovery is not applicable for granting patent. However, over the time period, several amendments have been made for granting patent. One of the major breakthroughs in the history of IPR is the granting patent in the field of biotechnology. Biotechnology is the utilization of biological organisms through changes in different biological processes for product development or technological intervention to modify any biological processes for the benefit of the society. An IP rights in this field of technology has revolutionized the biotech industries in the world (Table 10.1).

10.3 PATENT AND BIOTECHNOLOGY

Classical biotechnology may be defined loosely as the production of usual products by living organisms. The definition given by the US Office of Technology Assessment state that "biotechnology includes any technique that uses living organisms (or parts of organisms) to make or modify products, to improve plant or animals or to develop microorganisms for specific use." Since beginning, biotechnology can be traced back to various stages of its development (Archana, 2013). The first phase of biotechnology was based on the traditional knowledge in various tribes for the preparation of fermented foods or their product thereof, medical distillates, etc.

TABLE 10.1 Historical Developments Made in the Field of IPR

Year	Patent grantee	Patent description
600 B.C	Robin Jacob	A patent was documented for "some kind of newfangled loaf" of bread
1421	Filippo Brunelleschi	Developed crane system carrying marble slab from the Carrara mountains
1441	Vinetean Act	The first patent law articulating the concept of intellectual property and enshrining the importance of protecting inventors' rights
1624	British statute of Monopoly	The royal court bestowed patents on well-established techniques or commodities (vinegar and playing cards, for instance).
1790	The US Patent Act	The First patent act was passed in US
1883	Paris Convention	Industrial property rights: Priority Rights and Unfair competition
1886	Berne Convention	Literary and Artistic Rights
1891	Madrid Agreement	International Registration of Mark
1958	Lisbon Agreement	Appellation of Origin
1961, 1978 & 1991	International Union for the Protection of New Varieties of Plants (UPOV)	Plant Variety Protection and Breeders' Rights
1967	World Intellectual Property Organization (WIPO)	Promotes IP protection through out the world
1983-93	FAO International Undertaking on Plant Genetic Resources (IUPGR)	Plant Breeders' Rights and Farmers' Rights
1993	Convention on Biological Diversity	Sustainable utilization of genetic resources; fair and equitable benefit sharing
1995	TRIPS Agreement	Plant Variety Protection, Integrated circuits, Trade secrets
2001	International Treaty on PGR for Food and Agriculture (IT-PRGFA)	Multilateral System for access and benefit sharing
2010	Nagoya Protocol of CBD	Access to Genetic Resources and the Fair and Equitable Sharing of Benefits Arising from their Utilization

The second phase of biotechnology started when micro-organisms were utilized for the production of industrial products. Most commonly, these included mass production of alcohol; fermentation of antibiotics; and development of classical vaccines for cholera, typhoid, yellow fever, etc. Louis Pasture patented several products related to the production of beer. The production of beer was patented by Pasture in France, England, Italy, and United States.

The third phase of biotechnology began in the 1970s with the two basic techniques of recombinant DNA technology (rDNA) and Hybridoma technology. These modern technologies are distinct from classical fermentation technology that helps us to change the biological processes in a desired direction. In rDNA technology, foreign genes are inserted into the genome of an organism to produce product of our desire; in the hybridoma technology, antibodies producing B-cells (a type of white blood cells) are fused with myeloma cells (immortal cancer B-cell) to produce a hybrid cell line called a hybridoma, which possesses the antibody-producing ability of the B-cell and immortality and reproductivity of the myeloma cell.

Recently, the fourth phase of biotechnology can be related to advancement made in interdisciplinary techniques like bioinformatics, information technology and nanotechnology. In bioinformatics, computer-based analysis of biological data is carried out to determine the biological information.

10.4 PATENTING ON MICROBIAL BIOTECHNOLOGY

Microbes are one of the top listed organisms where patents were granted in the field of biology. The story begins when Louis Pasteur received a patent for a process of fermenting beer. Besides beer, acetic acid fermentation and other foods, patents date from the early 1800s up to the 20th century patents on microbes produced through rDNA technology by AM Chakroborty. The development of rDNA technology, i.e., the controlled joining of DNA from different organisms has resulted in greatly increased understanding of the genetic and molecular basis of life. However, granting of patent related to biotechnology remains under great controversies. Prior to 1980, patents were not granted for such inventions, which were considered to be "products of nature" and not statutory subject matter. Patent applications were rejected if directed to living organisms *per se*, under the context of product of nature. However, patent protection was granted for many compositions containing living things e.g., sterility test devices containing living microbial spores, food yeast compositions, vaccines containing attenuated bacteria, milky spore insecticides, and various dairy products (Archana, 2013). Louis Pasture received patent (US Patent No. 141072) on July 22, 1973 entitled "Improvement of manufacture of Beer and yeast" claiming the method of obtaining pure yeast by eliminating the organic germs of disease from a brewer's yeast. Yeast obtained herewith was free from organic germs of disease, and was considered as an article of manufacture as per US patent laws. In 1980, AM Chakraborty got the patent on genetically modified *Pseudomonas* bacterium that degrades oil spills. Since then, the patenting on microbial biotechnology has got new dimension called patenting microbial biotechnology.

The story started when *Pseudomonas* bacterium was manipulated to contain four plasmids that control the breakdown of hydrocarbon present in oil spill by AM Chakraborty. Initially, the patent application was rejected on two grounds:

1. microorganisms are "products of nature;" and
2. as living things, microbes are not patentable subject matter.

The case was heard by the US Supreme Court, which in a 5-4 ruling, held that a live, human-made microorganism, which is not found in nature, is either manufactured or is composition of matter and thus a patentable subject matter. Following the decision of USPTO on Chakrabarty case, the Japan Patent Office (JPO) and European patent office (EPO) started granting patents on microbes.

10.5 PATENTS ON ANIMAL BIOTECHNOLOGY

The gene causing cancer in mammals was isolated and injected into fertilized eggs of mouse by Dr. Philip Leder, geneticist at Harvard Medical School, and Dr. Timothy A. Stewart, a former Harvard researcher who later became senior scientist at Genentech Inc. This genetically modified mouse is called Harvard Mouse and is suitable for studying how gene causes cancer in the mammals. The first animal patent was issued in April 1988 to Harvard University for the development of genetically engineered mouse

contained a cancer-causing gene (U.S. 4,736,866). Specifically, the patent covers "a transgenic non-human eukaryotic animal (preferably a rodent such as a mouse) whose germ cells and somatic cells contain an activated onco gene sequence introduced into the animal which increases the probability of the development of neoplasm (particularly malignant tumour) in the animal." After that many patents have been awarded on inventions related to transgenic mice as model specific to ulcers, photo Parkinson's syndrome, inflammation, sickle cell anemia, Alzheimer's disease, HIV infection, cutaneous melanoma, leukemia, thrombocytopenia, etc.

10.6 PATENTING ON PLANT BIOTECHNOLOGY

Genetically modified (GM) crops are produced through the introduction of foreign genes from other organisms. The first GM crop is "Flavr-Savr" tomato produced by delayed ripening in 1994 by Calgene, a California-based company. In many countries including India, plant variety cannot be patented. In USA, plant variety is a patentable subject matter either through plant patent, plant variety protection act, or utility patent. The US Plant Patent Act was adopted in 1930 that covers only asexually propagated crops, e.g., apple, grapes, rose, etc. It excludes seed-propagated crops and tuber-propagated crops like potato. Similarly, an act was passed in Europe for the protection of plant through patent. Later, in 1961, when Union for the Protection of New Plant Varieties (UPOV) was established to cover seed-propagated crop, the US Plant Variety Protection Act was passed in 1970. New plant variety propagated through seeds could be patented, but initially, this law excluded F_1 hybrids and cucumber, carrot, celery, okra, tomatoes, and pepper. But these were included in the Plant Variety Protection Act in the subsequent year. There was provision for breeders' exemption and farmer's exemption, where the breeders can use patented plant variety for the development of new plant varieties and farmers' were allowed to save the seeds from previous year planting of a patentable variety and also allowed to sell the saved seed of that variety to other farmers. But after the amendments in 1994, breeders were not allowed to patent a variety that was essentially derived from the already existing patented variety. Farmers were not allowed to sell the saved seed to other farmers but can save seeds to grow next year.

In the year 1985 the United States declared that Utility Patent Act of 1790 were also applied to patent protection for plants. It mostly covers the genetically modified plant varieties. Utility patent covers processes to genetically engineer a plant or insert genes or product of the genes produced through genetic engineering. There is no provision of breeders' or farmers' exemption. Farmers are not allowed to save their own farm seed for next year planting or to sell out. With the development of genetically modified (GM) crops, the strong IP regulation is essentiality for developed and developing nations. In India, transgenic plant varieties can be protected as essentially derived varieties (EDVs) under Protection of Plant Varieties and Farmers' Right Act 2001.

10.7 WHAT ARE PATENTABLE IN THE FIELD OF BIOTECHNOLOGY?

The nature of any IP law is territory-based; this means that provisions of any IP laws vary from nation to nation and what is patentable in one state may not be patentable in another state. One example is plant cannot be patented in India but can be patented in USA. Based on economic and social structure of the nations, they are independent to regulate the nature of IP laws. However, the Article 27.3(b) under TRIPS agreement addresses one of the most controversial issues related to whether plant and animal inventions should be covered by patents, and how to protect new plant varieties. This describes that members may exclude from patentability of plants and animals and essential biological processes but members may protect microorganisms, certain biotechnological processes, and plant varieties either by patents or by an effective *sui generis* system or both.

According to the Indian Patent Act 1970, any subject needs to pass the criteria for patentability. The subject must be new, should have an inventive step and industrial applicability, and must not follow the criteria set under section 3 of Patent Act. The following are excluded under section 3 of patent act:

1. The invention should not primarily or intended to use contrary to *public order* or morality or should not cause any serious prejudice to plants, animals, or human beings or to the environment.

For example, genetic modification that results in suffering to any animals or harmful to environment is not patentable.

2. Any subject that is mere discovery rather than invention is not patentable. This means that anything isolated from nature is not a patentable subject matter. Thus, microbes isolated from nature or DNA, RNA, or proteins are not patentable.

In 2005, recombinant DNA molecules or plasmids were considered under patentable subject matter provided there must have sufficient human intervention. However, transgenic animals or plants or their parts like seeds, vegetative parts, varieties and their genes were excluded from patentability. A novel genetically modified gene or amino acid sequence produced through an inventive step and has industrial applicability is patentable. As a result, gene sequence, the method of expression of such sequence, and antibody made against any protein are patentable in India.

According to the European Patent rule, isolated genes, proteins, enzymes, antibodies, viruses, bacteria, and stem cells including transgenic plants or animal are patentable, provided they are not suffering without any medical benefit. Isolated DNA sequences are not patentable in USA, but cDNA or any modification to the DNA sequence or protein is a patentable subjected matter because they do not exist in nature.

10.8 WHAT IS A GENE PATENT?

Gene patents, more specifically comprise patent claims to nucleotide sequences such as genes, plasmids, and probes, which fit the criteria of patentability like novelty, utility, and industrial importance. In US, patents have been issued on whole genes whose functions are not yet known (novel genes). Also, inventors seek patents on partial gene sequences. Some patents have been granted for these. In fact, by 2008, there were over 3 million genome-related patent applications that had been filed in the US alone. Several gene patent granted for therapeutic proteins like human insulin, mAbs, Herceptin®; transgenic plants, insect-resistant corn; and diagnostic probes for genetic diseases.

Further limitation on research could come from a US patent for the gene gun, one of the most common means for inserting genes into plants. It was issued to Cornell University, which licensed it to DuPont. Similarly, Monsanto holds a patent on the 35S promoter, a portion of DNA that is often inserted with a plant gene to promote its expression. If breeders cannot use such tools or need licenses to use them, it will be substantially more difficult and expensive for them to produce superior seed [ues.org/17-4/barton and berger, 2011].

10.9 GLOBAL SCENARIO ON PATENTABILITY OF GENES

Genes patents comprise claims over isolated gene, gene-constructs, plants transformed with the constructs, and also seed and progeny plants that are being routinely patented currently. The first gene patent (US 4,447,538) claimed a recombinant DNA transfer vector containing the *Chorionic Somatomammotropin* gene appeared after 1982. The 1980 ruling *Diamond vs. Chakrabarty* paved the way for the patenting of genes by declaring genetically engineered bacteria to be a patentable material. The 1980s saw a rise in gene patents to 4,459. By 1999, the numbers had increased to 26,401 [www.cas.org]. The gene patent of the Xa21 Kinase gene by the University of California, which makes grains resistant to disease, to protect IRRI's right to use the gene. The patents for *Bacillus thuringiensis* (Bt) technology containing bacterial genes encoding toxic proteins with insecticidal property lies with four or more different companies.

10.9.1 UNITED STATES

Naturally occurring biological substances can be patented if they are sufficiently "isolated" from their naturally occurring states. Prominent historical examples of such patents include those on adrenaline, insulin, vitamin B_{12}, and various genes. The *Myriad Genetics* case resulted in a ruling that because isolated DNA is not markedly different from DNA as it exists in nature, it constitutes an un-patentable subject matter. While man-made genetic constructs such as cDNA are patentable, isolated DNA without further alteration or manipulation is not patent-eligible. United States has been patenting chemical compositions based upon human products for over 100 years. The first patent for a human product was granted on March 20, 1906, for a purified form of adrenaline. It was challenged and upheld in *Parke-Davis vs. Mulford.* Judge Hand argued that natural substances when they are purified are more useful than the original natural substances. The 1970s marked the first time when scientists patented methods on their biotechnological inventions with recombinant DNA. In 1980, the US Supreme Court, in *Diamond vs. Chakrabarty case*, upheld the first patent on a newly created living organism, a bacterium for digesting crude oil in oil spills. The patent examiner for the United States Patent and Trademark Office had rejected the patent of a living organism, but Chakrabarty appealed. As a rule, raw natural material is generally rejected for patent approval by the USPTO. The Court ruled that as long as the organism is truly "man-made," such as through genetic engineering, then it is patentable. Because the DNA of Chakrabarty's organism was modified, it was patentable. Since that 1980 court case, there has been much patenting of genetically modified organisms. This includes bacteria, viruses, seeds, plants, cells, and even non-human animals. Isolated and manipulated cells-even human cells-can also be patented (https://en.wikipedia.org).

10.9.2 EUROPE

European Union directive (the *Biotech Directive,*98/44/EC) allows for the patenting of natural biological products, including gene sequences, as long as they are "isolated from [their] natural environment or produced by means of a technical process."

10.9.3 JAPAN

Japan Patent Office (JPO) guidelines for biology-related inventions is similar to that of EPO. The Japanese Patent Act requires that patented inventions be "industrially applicable," i.e., they must have market or commercial potential.

10.9.4 CHINA

In China, the development of technology of hybrid rice has huge impact on country's rice production. The National Seed Corporation of China held patents for these technologies, but these patents are no longer in force; the company patented certain aspects of the technology in the United States. These patents deny breeders access to research tools that could be useful in developing new varieties of many crops. Patents have also been granted on other ways to produce hybrid seed (Barton and Berger, 2001).

10.9.5 INDIA

The Patents (Amendment) Bill 1999 allows patent on seeds and plants by introducing amendments in the exemptions covered by Section 3 of Chapter 2, Indian Patent Act 1970. These amendments enable patents on processes to modify plants to render them free of disease and to increase their economic value; micro-organisms including parts of seeds varieties and species as well as plants and animals produced through genetic engineering (methods other than essentially biological processes). The clause (ii) of Section 4 is a "patent on life" clause because it omits the word "plants" from Section 3(I) of the Indian Patent Act 1970; hence, patents can be granted for processes for "treatment of plants to render them free of disease or increase their economic value or that of their products." Because genetic engineering allows genes to be introduced in plants for pest and disease resistance as in the case of Bt. cotton where a gene from a bacterial species was added to cotton. Such a method of agriculture is now patentable under 3(1), though it is excluded under 3 (h). The amendment has thus undone an exemption of patents in agriculture, which is vital for the rights and survival of farmers. The exemption under 3 (h) for patents in agriculture has been undone by amended 3(I). While the Patents Act 1970 in India prohibits patenting naturally occurring substances, patents covering genetic material and nucleotide sequences have been granted. For non-naturally occurring genes, once their functions are delineated or utility specified, they become patentable. In 2005, biotechnological issues have been annexed and patentability of recombinant DNA and plasmid involving huge human intervention granted. The guideline for the examination of Biotechnology Application for Patent, 2013 states that sequences isolated directly from nature are not patentable.

10.10 PERSPECTIVE ON GENE PATENTING IN AGRICULTURE

10.10.1 GENE PATENTING IN AGRICULTURE

One of the earliest examples of an agricultural genetic patent was "Flavr Savr" tomato with long shelf-life. Calgene got a patent on the technique of flipping a genetic sequence (polygalacturonase, or PG). The patent protected the production of "anti-sense" genes. So, in 1996, Monsanto purchased the rights and patents to the Flavr Savr. They had no intention of marketing the tomatoes but saw the value in Calgene's broad patent on anti-sense genes. The patent could apply to any gene-spliced food plant that used the technique, and Monsanto would charge licensing fee for the technique. That same year, 1996, Monsanto received a patent on the genes that allowed their Bt crop varieties. After that, they make millions off of their patents for genetic materials. India Patent office granted a patent for the invention titled "Synthetic gene encoding a chimeric δ-endotoxin of *Bacillus thurigiensis*" (Patent No.: 237912). In India, the Patent office granted a patent for the invention titled "Synthetic gene encoding a cry1Fa1 S-endotoxin of *Bacillus thuringiensis*" (Patent No.:242768). Several patents related to agriculture have been filed by CSIR, ICAR, DBT, and state universities. For example, complete patent application for invention titled "Novel trichome specific promoter" was filed on October 30, 2009 (no. 2251/DEL/2009). Another complete patent application for invention titled "Rice polynucleotide associated with blast resistance and uses thereof" was filed on February 4, 2010 (no. 241/DEL/2010).

The varieties such as IR-16 and IR-64 were developed under donor funding at IRRI during Green Revolution. IRRI freely provided these new varieties and other innovative breeding materials to many national research centers in the major East Asian nations. In contrast to such noble initiative, the global patent searches show that several MNCs involved in agricultural research are seeking to protect their IP positions in developing nations, so that in future they charge for their breeding material.

10.10.2 GOLDEN RICE AND GENE PATENT ISSUES

Golden rice is a noble example of public-private partnership for humanitarian purpose. This technology is freely accessible to researchers or anyone who need it. The Swiss and German scientists developed this GM rice rich in vitamin A containing a piece of DNA isolated from the bacterium *Erwinia uredovora* and daffodils that gives the rain a golden yellow hue. The rice is golden because a biosynthetic pathway is engineered to produce gold-colored beta-carotene, a precursor to vitamin A. Golden rice involves patents for more than 20 different entities held by Professor Ingo Potrykus, (Swiss Federal Institute of Technology Zurich), and Professor Peter Beyer (University of Freiburg), Syngenta, Bayer AG, Monsanto Co, Orynova BV, and Zeneca Mogen BV. These companies provided access to the required technologies free of charge for humanitarian purposes. Like other GMOs, this rice is also facing strong opposition by anti-GMOs groups such as Greenpeace, opposing the cultivation of golden rice, claiming it will open the door to more widespread use of GMOs in agriculture.

10.10.3 CONTROVERSIES RELATED TO GENE PATENT

Inventions do not move from the laboratory to the marketplace without a huge investment of money, time, and effort. In 2008, the Canadian activist organization, the ETC Group (Erosion, Technology and Concentration) estimated that corporations like BASF, Monsanto, Bayer, Syngenta, and Dupont had filed over 530 patent applications for genes in plants that could withstand changes caused by global warming. If granted, these patents would give the companies monopolies on GMOs that could withstand drought, heat, cold, floods, saline soils, and more, far into the future [www.livinghistoryfarm.org]. Some critics say patents could hamper the development of useful plants or drugs by others because of the high cost of getting the right to research patented genes.

Other critics say that granting patents on partial or uncharacterized DNA sequences will reward those who make routine discoveries but penalize those who determine the biological functions of the DNA. Some are concerned that there will be multiple patents issued on the same gene, first as a short string, then as a longer one, then the full gene, and then the gene and its function. Multiple patents increase the cost of any new plant, animal, or drug.

10.11 IMPACT OF GENE PATENTING ON AGRICULTURAL RESEARCH AND DEVELOPMENT

Many big private firms like Aventis, Dow Chemical, DuPont, Monsanto, and Syngenta are now holding a large portion of patents in agriculture. One of the apprehensions is that patent granted for genes will give monopoly to institutions and private companies; as a consequence, a huge sum of money will be charged by them as license fee for use by others. Monsanto's patent application on aspects of climate resilient traits of cold tolerance, salt tolerance, and drought tolerance that our farmers have evolved over thousands of years through applying their knowledge of breeding. Monsanto owns more than 1500 patents on climate resilient crops. Along the coastal areas, farmers have evolved flood-tolerant and salt-tolerant varieties of rice such as "Bhundi," "Kalambank," "Lunabakada," "Sankarchin," "Nalidhulia," "Ravana," "Seulapuni," and "Dhosarakhuda." With these very broad patents, corporations like Monsanto can prevent access to climate resilient seeds after climate disasters, because a patent is an exclusive right to produce, distribute, and sell the patented product [http://www.navdanya.org].

Developing nations are mostly protecting intellectual property-related works on crops by PBRs in the line with 1994 TRIPS agreement, which is less stringent than patenting rights. A new initiative was carried out by International Rice Genome Sequence Working Group, which includes Japan, Korea, China, the United States, the European Union, and the Rockefeller Foundation. All the sequence information will be publicly available, with no patenting of these sequences. A similar move by private firms like Monsanto, Syngenta, and Myriad Genetics, who completed rice genomic sequencing of their own and promised to provide the information for public research and developmental work in developing nations, is a welcome step that can impact globally the future research program of public or private sectors.

10.11.1 LANDMARK US COURT VERDICT ON GENE PATENT

The landmark judgment by US court in the case of Myriad Genetics, the patents of seven human genes including BRAC1 and BRAC2 granted in early 2010, invalidated by court classifying these genes natural and unpatentable, brings hope against gene patent in agriculture domain as well. However, the court held that cDNA, a synthetic form of DNA, may get patent protection.

10.11.2 ETHICAL AND LEGAL IMPLICATION

The patentability of genes has been a controversial issue that impacts ethics, legality, and scientific research and development globally. The idea of patenting an element of life is immoral: owning or treating genetic material as a property is a concern. There is also apprehension that allowing patents of genetic material could lead to monopolies exhibiting amoral behavior in healthcare and other industries. As synthetic biology is a relatively new field with unprecedented possibilities, there is a great debate surrounding gene patenting in moral, practical, and legal domains. On the legal end, genetic patents are currently on the hot seat in the courtrooms with some patents being upheld, and others not. Many academics feel that the legal patent requirements of "utility," "non-obviousness," and "sufficiently isolated or transformed" are not being appropriately met. A major concern is that the use of patented materials and processes will be very expensive and can impact the access of agricultural institutes and university researchers to genetic materials.

India approved commercial cultivation of Bt cotton in 2002. It was developed by Mahyco-Monsanto Biotech. The Indian government has woken up to the fact that Monsanto's bacterial gene that creates resistance to boll worms in cotton plants went off patent three years ago. It turns out that Monsanto never held the patent on the gene in India. Moreover, the US patent it holds expired in 2012. It has been suggested to the ICAR that the gene be incorporated in re-sowable straight line varieties and supplied cheaper to farmers than Bt cotton hybrids.

10.12 CONCLUSION

The patent, in essence, is a safeguard against copying, selling, and making profit out of others innovations involving huge investment of money, time, and human efforts. At the same time, it hinders the progress of scientific research and development by debarring the access to scientific data and research material, which otherwise could be of use for further development. Taken together, pros and cons of patenting genes with agricultural perspective, it should be ethically guided and human-welfare oriented rather than targeted toward making profit.

KEYWORDS

- **agriculture**
- **gene patentability**
- **genetic material**
- **transgenics**

REFERENCES

Archana, K., (2013). Do we need patent protection in biotechnology inventions? *International Journal of Scientific and Research Publications, 3*(4), 1–5.

Barton, J. H., & Berger, P., (2001). Patenting agriculture, *Issues in Science and Technology, 17*(4).

Dewan, M., (2011). IPR Protection in Agriculture: An overview, *Journal of Intellectual Property Rights*, *16*, 131.

Fischer, K. S., Barton, J., Khush, G. S., Leung, H., & Cantrell, R., (2000). Collaborations in rice, *Science*, *290*, 279–280.

Kochhar, S., (2011). Analysis of opportunities and challenges in IPR and agriculture in the Indian context. *Journal of Intellectual Property Rights, 16*, 69.

Kryder, R. D., Stanley, P. K., & Anatole, F. K., (2000). The intellectual and technical property components of pro-vitamin A rice (GoldenRice™): *A Preliminary Freedom-To-Operate Review,* ISAAA Brief No. *20.*

Mittal, R., & Singh, G., (2005). Patenting activities in agriculture from India. Jour*nal of Intellectual Property Right, 10*, 31.

Thomas, S., et al., (1999). Plant DNA patents in the hands of a few. *Nature*, *399*, 405–406.

Wright, B., (2000). Intellectual property rights: Challenges and international research collaboration in agricultural biotechnology [In: *Agricultural Biotechnology in Developing Countries: Toward Optimizing the Benefits for the Poor*, eds. Qiam, M., Krattinger, A. F., & Von Braun, J.], Kluwer Academic Publishers, the Netherlands, pp. 289–314.

CHAPTER 11

AWARENESS GENERATION, SCOPE, AND SCENARIO OF INTELLECTUAL PROPERTY RIGHTS IN AGRICULTURE

AJEET KUMAR

Regional Research Station, Madhopur, West Champaran–845454, Dr. Rajendra Prasad Central Agricultural University, Bihar, Pusa (Samastipur), India, E-mail: ajeetrpcau@gmail.com, ajeetrau@gmail.com

CONTENTS

ABSTRACT

Intellectual Property Rights (IPRs) may be broadly defined as "legal rights governing the use of creations of the human mind." This term covers a bundle of rights, each with not only different scope and duration but with a different purpose and effect. All IPRs generally exclude third persons from commercially exploiting protected subject matter without the explicit authorization of the right holder for specified duration of time. This enables IPR owners to use or disclose their creations without the fear of loss of control over their use, which ultimately helps in their dissemination. It is generally believed that IPRs encourage creative and inventive activity and make for orderly marketing of proprietary goods and services. Protection against unfair competition is the underlying philosophy for all IPRs, although there are some specific rules in international intellectual property law targeted toward these issues. IPRs are limited to a defined territory and have historically been attuned to the circumstances and needs of different jurisdiction.

There has been a revolution in science in agriculture over the last 10 years, primarily in the area of biotechnology. At the same time, there have been substantial changes in the application of IPR to scientific discovery in the life sciences. In addition to the technical and legal shifts, there has also been a move toward greater globalization of trade. There are concerns with regard to access of technology by developing countries; there are concerns regarding the rights to use of germplasm, the basic building blocks for genetic improvement; and there are concerns that technology is perceived to be controlled by a limited number of large corporate entities. This chapter describes some of the key technology tools that are used in plant improvement; it also describes the basic IPR tools that are used to protect ownership rights. It is noted that the IPR tools vary from jurisdiction to jurisdiction, but there are common platforms such as the Paris Convention that govern the international implementation of these rules. There are many fundamental beliefs. Of course, the subject matter is complex and involves analysis that incorporates science, law, and ethics. There is no right answer on many of these issues; only the possibility of compromise and consensus, such that fairness and equity are seen as a part of the outcomes.

11.1 INTRODUCTION

Intellectual property rights (IPRs) can be broadly defined as legal rights established over creative or inventive ideas. Intellectual property is intangible, because it is an idea or invention. The legal mechanisms of patents, copyrights, trade secrets, and trademarks are used to protect such intangible property. Keep in mind that some contract mechanisms such as licenses or material contract agreements (MTAs) have the effect of conveying ownership rights over materials. A basic understanding of these mechanisms is essential for anyone whose research may lead to an invention and for research administrators who must deal with intellectual property issues, both for acquisition and deployment.

Such legal rights generally allow right holders to exclude the unauthorized commercial use of their creations/inventions by third persons. The rationale for the establishment of a legal framework on IPRs is that it is a signal to society that creative and inventive ideas will be rewarded. This does not mean that there is no other way of rewarding such ideas or that this system is absolutely necessary, even less sufficient, to reward inventiveness or creativity. Nevertheless, it would be difficult to deny that IPRs do have a role to play in setting up of any such reward system.

The efforts are being made by many institution, as there is a critical need to increase IPR literacy, not only in terms of laws, rules, and procedures but also in terms of increasing the awareness on the long-term benefits for the country, particularly from increased domestic research and development and productivity.

IPRs have never been more economically and politically important or controversial than they are today. Patents, copyrights, trademarks, and geographical indications (GIs) are frequently mentioned in discussions and debates on such diverse topics as human rights, public health, agriculture, education, trade, industrial policy, biodiversity management, biotechnology, information technology, the entertainment and media industries, and increasingly the widening gap between the income levels of the developed countries and the developing, and especially least-developed, countries. There is no doubt that an understanding of IPRs is indispensible in all areas of human development.

"Intellectual Property shall include the rights relating to; literary, artistic and scientific works, performances of performing artists, inventions in all fields of human endeavor, scientific discoveries, Industrial designs, trademarks, service marks and commercial names and designations, protection against unfair competition, and all other rights resulting from intellectual activity in the industrial, scientific, literary or artistic fields" (WIPO Convention).

IP is usually divided into two categories: copyright and industrial property. Industrial property rights include patents for inventions, trademarks, industrial designs, plant variety rights, and of course, GIs. The holder of IPR has a legally recognized capacity to authorize and/or prevent others from acting in certain ways with respect to his/her intellectual property.

11.2 NATURE AND KINDS OF INTELLECTUAL PROPERTY

- Creation of human mind (Intellect)
- Intangible property
- Exclusive rights given by statutes

- Attended with limitations and exceptions
- Time-bound
- Territorial (Patent terms vary considerably from country to country)
- Different kinds of properties are-
- Movable Property: Car, Pen, Furniture, Dress
- Immovable Property: Land, Building
- Intellectual Property: Literary works, inventions

Important IP laws that prevail in most of the countries are patents, copyright and related rights, trademarks, trade secrets, industrial designs, industrial property, GIs, layout designs/topographies, and protection of new plant varieties.

11.3 PATENT PROTECTION

Patents provide inventors with legal rights to prevent others from using, selling, or importing their inventions for a fixed period. Applicants for a patent must satisfy a national patent-issuing authority that the invention described in the application is new, susceptible of industrial application (or merely "useful" in the United States), and that its creation involved an inventive step or would be unobvious to someone skilled in the art represented by the claimed invention.

There are three basic statutory types of patents:

- A utility patent is the type with which most people are familiar, and it is granted for any new and useful process, machine, manufacture, or composition of matter or any new or useful improvement thereof. In simple terms, it has to be useful.
- A design patent protects a new, original, and ornamental design for an article of manufacture
- A plant patent protects a new and distinct, asexually reproduced variety of a plant.
- Maintenance fees on utility patents must be paid at 4, 7, and 11 years after the date of issue or the patent will expire. Once a patent expires, the invention is in the public domain, and anyone may use it without authorization from the patent holder.

- A patent is an agreement between the Government and the inventor.
- A patent is an exclusive right granted for an invention, which is a product or a process that provides a new way of doing something or offers a new technical solution to a problem.
- The limited monopoly right granted by the state enables an inventor to prohibit another person from manufacturing, using or selling the patented product or from using the patented process, without permission.
- Period of Patents – 20 years.

In several countries, patent legislations were established as early as the 19th century. In these legislations, it was specified that patent protection would be offered only to certain categories of inventions. The general patentability requirements were similar at the basic level in most of the national patent acts, which required inventions to be novel and industrially applicable. The nonobviousness or inventive step requirement was established later on initially by case law in the mid-19th century and subsequently by codification. The 1930 US act introduced a special kind of plant patent for vegetatively propagated materials, but in the US, standard utility patents can also now be granted on plant varieties. Therefore, in the US, there exist today a dual system of protection; plants can be protected by a *sui generis* plant protection Act as well as by the Patent Act. Ever since the first generations patent acts, patent protection for plants has always been questioned. Between 1790 and 1970, plants patent protection was denied on the ground of noncompliance of the plant inventions with the legal requirements of patentability. In Europe and the US, the legal requirements that were not fulfilled by the plant inventions were conceptions, novelty, inventive step (or nonobviousness), industrial applicability (or utility) and adequate disclosure.

In India, patents are available for new process in respect of agriculture but not to all products per se. Patens may be obtained in agriculture for process related to agrochemicals, growth promoters and regulators, vaccines, drugs, hides and wool, diary technology, food technology, fuel and biogas production, bioreactors, standardizations of various laboratory protocols, environment management, and the like. However, there are certain restrictions under section [3(h, 3(i)) & 3 (j)] of the Patents Act 1970 related to grant of agricultural inventions.

Conditions of patentability are: (a) novelty – invention not known to public prior to claim by inventor; (b) nonobviousness (inventive step) – invention would not be obvious to a person with ordinary skill in the art; and (c) industrial application – invention can be made or used in any useful, practical activity as distinct from purely intellectual or esthetic one.

Features of patents are patents granted by national patent offices after publication and substantial examination of the applications; they are valid within the territorial limits of the country; international patent can also apply for more than one country. In India, provisions exist for pre-grant and post-grant opposition by others.

11.4 COPYRIGHTS

In contrast to a patent, which protects an idea and its implementation, copyright protects the expression of an idea, not the idea itself. Such expression must be in some retrievable form such as handwriting, set in type, or recorded on magnetic tape or other storage medium.

Copyright covers the expression in literary or musical works, computer programs, video or motion pictures, sound recording, photographs, and sculpture. Unlike patents, copyrights automatically come into being when the idea is fixed in a tangible medium of expression. Copyright gives authors legal protection for various kinds of literary and artistic work. Copyright law protects authors by granting them exclusive rights to sell copies of their work in whatever tangible form (printed publication, sound recording, film, etc.) is being used to convey their creative expressions to the public. In theory, legal protection covers the expression of the ideas contained, not the ideas themselves. In practice, information may also be protected, as when copyright is extended to cover new types of work such as software programs and databases. The right usually lasts for the life of the author plus 50 years, though in some jurisdiction this has been extended recently to 70 years.

11.5 TRADEMARK

A trademark is a word, name, symbol, or device used by a person or legal entity to identify their goods and distinguish them from others. Commercial logos are common examples of trademarks. Trademark rights can be asserted by using the familiar trademark indicator ™ in association with particular goods or services. Any unauthorized use of mark identical (or confusingly similar) to a valid trademark is prohibited. Protection of trademarks does not have a time limit, provided they are used and renewed periodically.

Trademarks are marketing tools used to support a company's claim that its products or services are authentic or distinctive compared with similar products or services of competitors. They usually consist of a distinctive design, word, or series of words placed on a product label. Normally trademarks can be renewed indefinitely, though this is likely to be subject to continued use. The trademark owner has the exclusive right to prevent third parties from using identical or similar marks in the sale of the same classes of goods or services, and thereby confuse customers.

Trademark can be sign, words, letters, numbers, drawings, pictures, emblem, colors or combination of colors, shape of goods, graphic representation or packaging, or any combination of the above as applied to an article or a product.

Main features of trademarks are:

- Trademarks are registered by national trademark registries and are valid in that country.
- Registration is made after examination and publication.
- Period of registration is for 10 years but can be renewed indefinitely.

Different kinds of Trademarks are:

- **Service Marks:** Service Marks include banking, education, finance, insurance, real estate, entertainment, repairs, transport, conveying news and information, advertising, etc.
- **Certification Trademark:** Certified by the Proprietor as having characteristics like geographical origin, ingredients, quality, e.g., AGMARK, WOOLMARK. Certification mark cannot be used as a trademark. It certifies that the goods on which it is applied are made

of 100% wool. It is registered in 140 countries and licensed to the companies which assure that they will comply with the strict standards set out by the Woolmark company, the owner of the mark

- **Collective Marks:** It is a mark that distinguishes the goods or services of members of association from marks of other undertakings. Association of persons can own a collective mark. It could be manufacturers, producers, suppliers, traders, or other profession bodies like institute of chartered accountants, test cricketers' association, etc.
- **Well-known Marks:** For example, *Coca-Cola* for soft drink, *Toblerone* (Triangular-shaped chocolates).

11.6 GEOGRAPHICAL INDICATION OF GOODS

The GI is an indication that identifies goods as agricultural goods, natural goods or manufactured goods as originating, or manufactured in the territory of country, or a region or locality in that territory, where a given quality, reputation, or other characteristic of such goods is essentially attributable to its geographical origin.

Article 22 of the World Trade Organization's (WTO's) Agreement on Trade-Related Aspects of Intellectual Property Rights (TRIPS Agreement) (1994) defines GIs as "*indications which identify a good as originating in the territory of a Member, or a region or locality in that territory, where a given quality, reputation or other characteristic of the good is essentially attributable to its geographical origin.*" A geographical name can therefore operate as a GI once a given quality, reputation, or other characteristic of the product using the name is essentially attributable to its geographical origin.

A GI is a sign used on a product to denote its origin. For example, "Bordeaux" is a GI for wine originating from the region of Bordeaux in the south of France, where it has been produced since the 8th century. Similarly, "Tequila" is a GI for liquor originating from the town of Tequila in the state of Jalisco, Mexico, where the liquor has been produced for over 200 years. Other examples of GIs include "Darjeeling Tea" "Champagne" "Alphonso" and "Jamaica Blue Mountain Coffee."

A GI can be a geographical place name (e.g., "Champagne"), a symbol (e.g., a picture of the Eiffel Tower, the Statue of Liberty, and an orange tree), and the outline of a geographical area or anything else capable of identifying the source of a product.

11.7 TRADE SECRETS

A trade secret is any formula, pattern, device, process, tool, mechanism, compound, etc., of value to its owner, which is not protected by a patent and is not known or accessible to others. As long as it is kept secret, the owner may obtain a great deal of commercial benefit.

Some inventions, data, and information cannot be protected by any of the available means of IPRs. Such information is held confidential as a trade secret. Trade secret can be an invention, idea, survey method, manufacturing process, experiment results, chemical formula, recipe, financial strategy, client database, etc.

When Trade Secrets are preferred?

- When invention is not patentable;
- Patent protection is limited to 20 years, when secret can be kept beyond that period;
- When cost of patent protection is prohibitive;
- How to guard Trade Secret?
- Restricting number of people having access to secret information;
- Signing confidentiality agreements with business partners and employees;
- Using protective techniques like digital data security tools and restricting entry into area where trade secret is worked or held;
- National legislations provide protection in form of injunction and damages if secret information is illegally acquired or used.

11.8 INDUSTRIAL DESIGN

Design' means only the features of shape, configuration, pattern, ornament or composition of lines or colors *applied* to any article whether in two dimensional or three dimensional or in both forms, by any industrial process or means, whether manual, mechanical or chemical, separate or combined, which in the finished article *appeal to* and are judged solely by *the eye.*

11.9 LICENSING OF AN INTELLECTUAL PROPERTY

License is a permission granted by an IP owner to another person to use the IP on agreed terms and conditions, while he continues to retain ownership of the IP. Licensing creates an income source. It establishes a legal framework for transfer of technology to a wider group of researchers. It creates market presence for the technology or trademark.

Licensing conditions of IPRs:

- Owners of IP prefer to transfer technology through licensing agreements only.
- All rights or limited rights can be licensed.
- Can be exclusive or non-excusive or sole (owner and licensee).
- Most such agreements provide for royalty payment and non-transfer to a third party.
- Royalties can be upfront, part upfront, and part % per production/sale, only % per production/sale.
- The particular uses for which the IP can be used are also generally specified.

One needs to be careful about the Competition law.

11.10 THE ADVENT OF WORLD TRADE ORGANIZATION

The recognition of agriculture as a rule-bound enterprise of investment and profit making became obvious with its inclusion in the intergovernmental negotiations for the General Agreement on Tariffs and Trade (GATT) for the first time in the Uruguay Round (1986–1994). This round led to the establishment of the WTO in January 1995. Now, the WTO has at least half a dozen of intergovernmental agreements that directly affect agriculture. These are Agreements on Agriculture (AoA), Applications of Sanitary and Phyto-sanitary Measures (SPS), Technical Barriers to Trade (TBT), Anti-Dumping, Subsidies, and Countervailing Measures, Safeguards, and TRIPs. An understanding of the implications and the application of these agreements particularly the TRIPs, has become more important than ever before at every stage of planning, research, up scaling, and commercialization of agricultural technologies. The TRIPs Agreement is covered in an elaborate document—comprising 73 articles in 7 parts, namely, (i) General provisions and basic principles, (ii) Standards concerning availability, scope, and use of IPRs (iii) Enforcement of IPRs, (iv) Acquisition and maintenance of IPRs and related *inter partes* procedures, (v) Dispute prevention and settlement, (vi) Transitional arrangements, and (vii) Institutional arrangements. There are seven forms of IPRs recognized in the TRIPS Agreement. These include copyright and related rights, trademarks, GIs, industrial designs, patents, layout-designs (topographies) of integrated circuits, and protection of undisclosed information.

Presently, when the application of various forms of IPR in different areas of agriculture is implemented, we may face serious problems unless timely remedial measures are taken, awareness is spread, and also due emphasis is given on IPR literacy, higher education, and capacity building in the country. Following establishment of the international institutional mechanisms, such as, the Convention on Biological Diversity (CBD) and the WTO, and further, signing of International Treaty on Plant Genetic Resources for Food and Agriculture (ITPGRFA), the growing importance and the global scope of IPR in agriculture are well realized and recognized.

The IPR, is recognized as an asset and means of rewarding and harvesting the fruit of agricultural research and development. Recognition of intellectual property rights provides an effective means of protecting and rewarding innovators. This acts as a catalyst in technological and economic development. The essence of regulation of IPR by law is to balance private and public interests.

The CBD, objectives are to conserve biological diversity, to promote the sustainable use of its components, and to achieve fair and equitable sharing of the benefits arising out of the utilization of genetic resources. These objectives in some way or the other are affected by IPR. IPR are relevant for developing mechanisms in order to protect and in force control over information conceived which may be new crop or plant varieties, pharmaceuticals, herbicides and pesticides or new biotechnological products and processes. The recognition of the economic, ecological, and cultural importance of genetic resource and biological derived materials particularly after coming into force of the CBD has given place to a new scenario where IP protection has been given great importance.

11.11 SCOPE OF INTELLECTUAL PROPERTY RIGHT IN AGRICULTURE

Protection of all forms of IPR may be relevant in agriculture but its application has to be limited to the relevant domestic Acts in vogue. Hybrids in plants and animals may be protected *de facto* by not disclosing the parents, whereas protection for plant varieties may be availed by a *sui generis* system. The provision for Plant Variety Protection (PVP) made under the TRIPs Article 27.3(b), allows countries to provide such protection either through patent, or an effective *sui generis* PVP system or any combination of the two. Patents, in India, are so far available to new processes but not to all products *per se.*

In agriculture, patents may be obtained for processes related to agrochemicals, growth promoters and regulators, vaccines, drugs, hides and wool, dairy technology, food technology, fuel and biogas production, bioreactors, standardization of various laboratory protocols, environment management, etc. Copyrights and related rights, on the other hand, may be registered for databases, bioinformatics, genes and gene sequences, amino acid sequences, antibodies, etc. Application of industrial designs and the topographies of integrated circuits would be relevant, particularly in agricultural engineering. Nevertheless, in the days to come, IPR is likely to dominate the agricultural scenario irrespective of whether the technology in question is conventional or modern—biotechnology or information technology. Countries are required to enact/amend their domestic laws in accordance with the TRIPs Agreement and the between-country disputes have to be resolved at the WTO platform, according to its dispute settlement procedures. In this context, it is important to have in place well enacted laws corresponding to the different forms of IPR that not only keep in view the basic needs of the country but are also capable of tackling complexities, which might arise at the international level. In India, the Patents Act 1970 constituted the basic Principal Act on the subject. This Act hardly included innovations in agriculture under the patentable subject matter. In particular, it excluded methods of agriculture and horticulture as well as all innovations in the areas of treatment and protection of plants and animals from pestilence or those aimed at increasing their productivity and value of their produce.

Further, with such shifts in legal provisions and also national policies, increased private participation in agricultural R&D and far more public-private relationships, including both competition and cooperation in relevant areas, are imminent. Several legislative and institutional adjustments are being made in the TRIPs member country to gear up and face the challenges of globalization. Likewise, in India, enactment of new legislations on Protection of Plant Varieties and Farmers' Rights Act, 2001 and GIs of Goods (Registration and Protection) Act 1999, and amendment of Patents Act 1970 in 1999 and 2002 were done. The Biological Diversity Bill 2000 is in the process of enactment and revision of the Seeds Act 1966, is also receiving attention. The need to provide for protection in the areas specific to farm animal sector is also being realized. Effective implementation of IPR-related legislations in place and those in the offing is expected to have significant impact on the course of agricultural R&D in the country. Therefore, it is considered important to identify and develop various national policy options for addressing the emerging areas of IPR in agriculture, including the access to various protected technologies to the farmers, entrepreneurs and users. It is high time that a critical analysis of the system is undertaken for its strengths, weaknesses, opportunities, and threats (SWOT), to convert threats into opportunities and mitigate weaknesses through timely action.

The opportunity for IPR in agriculture should be improved along with mechanisms for enforcement, access to resources and technology, benefit sharing, equity, and justice in order to give durable effect to the national agricultural policy. A national commitment should be made in respect of effective institutional mechanisms and reforms, including the administrative, regulatory, legislative, and judicial reforms at all levels of government. Short- and medium-term fiscal plans must include elements of these reforms by providing resources to meet the costs of adjustment. Resources should be tied to commitments by successive central and state governments with much-needed incentive to innovators commensurate with the invention.

Keeping with the spirit of the intergovernmental agreements, the application of IPR must be effectively addressed by the national laws. Enhanced competitiveness together with increased production should be the target for various agricultural commodities having export prospects. These include high value commercial crops, animal breeds, spices, medicinal and aromatic plants, and products like milk, meat, fish, leather, and wool. Reduction in the cost of production at small farms should also be aimed so that exports become more competitive. Market-driven quality consciousness should be applied to lay far greater R&D emphasis and efforts to produce quality products that may fetch increased monetary returns per unit area.

11.12 SCENARIO OF INTELLECTUAL PROPERTY RIGHT

The importance of IPR in agriculture should categorize broadly in three issues.

11.12.1 PROTECTABLE SUBJECT MATTER IN AGRICULTURE

Patentable subject matter in agriculture and alternative forms of IPR were considered. It also covered what is not patentable and where protection can or cannot be granted under copyrights, designs, GIs of goods, trademarks, undisclosed information (trade secrets), PVP, etc.

11.12.2 TECHNICAL OPPORTUNITIES IN AGRICULTURE

IPR implications were observed on plant varieties, farmers' rights, biodiversity, and environment. It also covered the biotechnological opportunities. In addition, coverage was made in respect of technology transfer, bio-safety, institutional capacity building, and human resource development.

11.12.3 ENABLING CONDUCIVE ENVIRONMENT FOR ACCELERATED RESEARCH AND DEVELOPMENT AND GLOBAL COMPETITIVENESS IN AGRICULTURE

A broad range of issues like commercialization, competitiveness, safeguards, information management, indigenous and traditional knowledge (ITK), and orientation of research and development for technology development, transfer, trade, monitoring, and management in the national and international context should covered.

These matters are tackled as given in the following sections.

11.13 AWARENESS GENERATION FOR IPR

An intensive campaign should be launched for awareness generation in order to accept and apply IPR and to naturalize the IPR culture among all relevant sections of the society. Awareness should be spread in public to enable them to respond to various opportunities, challenges, and threats. Elaborate awareness tools like compact discs, documentary films, newspaper and advertisements should be developed and widely disseminated through mass media. In order to increase the IPR literacy in agriculture, compendia on IPR and technology transfer should be published for wide circulation. Such compendia should cover rules, procedures, forms, guidelines, other important tips and selected case studies on various provisions, admissibility and application, and infringement and remedies for various forms of IPR and also in comparison with other country laws (Figure 11.1).

11.14 IPR EDUCATION, TRAINING, AND HUMAN RESOURCE DEVELOPMENT

It is need of time to develop suitable curriculum right from the college level. In order to enhance the level of higher education for IPR in relation to agriculture, there must be at least one compulsory course at the undergraduate and postgraduate levels in all agricultural universities. Summer and winter schools and periodic training should be conducted in the country for scientists and technical staffs. Appropriate modules should

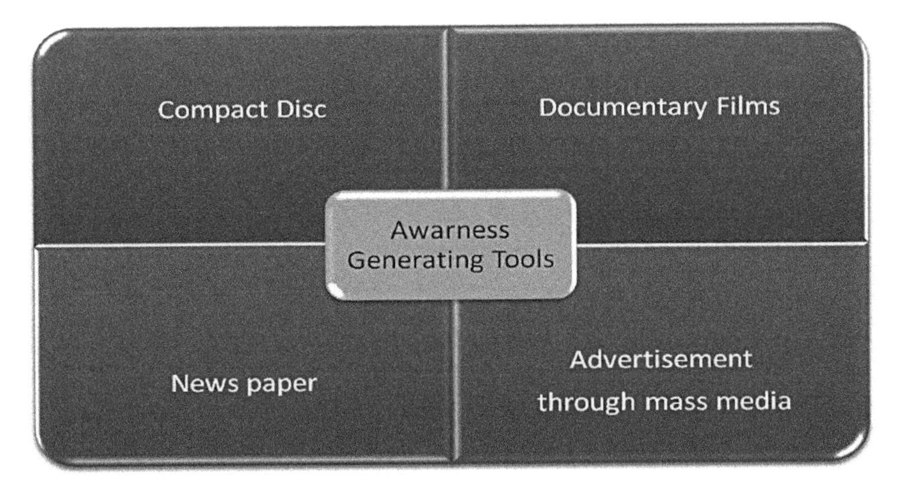

FIGURE 11.1 Different awareness generation tools for effective dissemination of technology in agriculture.

be developed for training and advanced orientation of scientists. Human resources should be developed and strengthened for efficient application of IPR in agriculture (Figure 11.2). Adequate funding should be provided to ascertain the much-needed promotion of human resource development.

11.15 HARMONIZATION OF THE IPR SYSTEM

Priority should be given to generation, evaluation, protection, and effective commercial utilization of tangible products of intellectual property in agriculture. A dynamic approach for IPR protection should be followed. Use of trademarks for brand development of agricultural products should be encouraged. Realizing the importance of biotechnology, hybrid technology, bio-control agents, bio-fertilizers, vaccines, diagnostics, improved implements, and machinery in agriculture should be fully harnessed. Recognizing the need to capitalize national resources to attain and sustain IPR advantages locally and globally. The area of IPR in agriculture should be addressed in conjunction with traditional rights and indigenous knowledge. Acknowledging the issues of IP protection based on our ITK is sensitive, a high priority and liberal financial allocation should be made for strengthening of traditional knowledge and resource databases. The

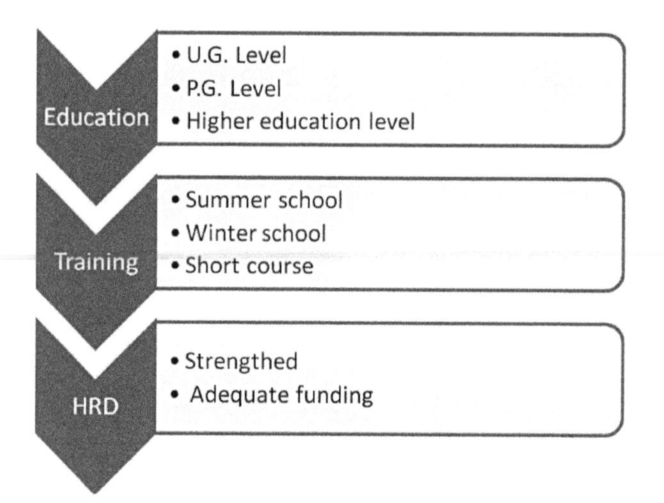

FIGURE 11.2 Different levels where learning on IP regulations can be effectively managed in agriculture.

intergovernmental commitment by developing countries to grant product patents in all fields of technology must be accorded to the development of competitive products, particularly in agrochemicals and biotechnology, in Indian agriculture, besides, further making suitable amendments in the Patents (Amendment) Act, 2002. The available strengths for animal genetic resources and generation of competitive technology in farm animals, poultry, and fish in the country, and also realizing that appropriate IP protection laws in this area are lacking. Steps should be initiated on the analogy of Protection of Plant Varieties and Farmers' Rights Act, 2001, so that in future animal and fish breeds/strains and also farmers' rights on these genetic resources are protected by law.

11.16 STRENGTHENING THE INSTITUTIONAL MECHANISM—LEGAL, REGULATORY, AND ADMINISTRATIVE

It is important to establish an IP regime that would provide confidence in and workability for the IPR in relation to agriculture; high priority should be accorded to the process of completing the required legislative provisions and also the notification, functioning, and strengthening of institutional mechanisms corresponding to various Acts, such as the respective Controllers, National Authorities, Tribunals, Registries, etc. The IPR Acts mainly relate to techno-legal matters, and their governance should be controlled by eminent scientists and the Tribunals should also have technical members. The National Authority on PPV&FR Act 2001 should have an eminent plant breeder as its chairperson. Enforcement of new Acts and Amendments related to IPR in agriculture should be hastened. The rules and procedures for the GI Act 1999 should be circulated on areas concerning agriculture. Enforcement of Patents (Amendment) Act 2002 should be done early to protect *inter alia* the wealth of agriculturally important microorganisms in the country. The designated repository should be equipped well and strengthened as per international standards. Similarly, enforcement of Amendment Acts related to copyrights and trademarks should be accorded to derive the best benefits. Development of related laws such as enactment of Biological Diversity Bill 2000 should also receive attention. Appropriate legal instruments related to conservation, maintenance, trade and sustainable utilization of animal genetic resources should be brought. Simplified regulatory procedures for relevant application of IP to genetic resources and equitable sharing of benefits should be developed. It is recommended that parallel laws like the Seeds Act should be strengthened in order to support effective implementation of *sui generis* system of protection. Similarly, Contract Law should be reviewed to strengthen the law on trade secret, and the law related to land ownership of small farm holders should also be strengthened to judiciously implement the farmers' rights. The institutional development and strengthening is complex in nature and time-consuming. Institutional culture should be developed, and short- and medium-term fiscal plans should include provision for resources. Facilities should be established and strengthened for the identification of research areas through patent search, literature survey, UPOV database search ,etc. Conflict-resolving services should be set up in the broader context. Inventors and innovators should be provided with their share commensurate with the worth of a commercialized invention whereas incentive should be given to all inventions in order to ensure an effective mechanism of intellectual property.

11.17 STRENGTHENING THE POLICY AREA

The codes and procedures for rewarding the concerned partners and stakeholder scientists should be developed. High priority should be given to support services in farm enterprises, extension, training, research, and quality control. Public interventions in agriculture should focus on market intelligence, technology forecasting and early warning systems. Market-led technologies should be developed, protected, and commercialized to harness greater returns on the investments. There is a strategic need to increase public investment. Agriculture is deregulated as a result of the ongoing reform process, and the lowest income groups should be continuously protected in accordance with clearly defined policy and directives. In the absence of proper legal framework, misuse, abuse, overexploitation, and non-judicious utilization of animal genetic resources is rampant; therefore, attention should be given for the same. There should be intergovernmental negotiations to address issues like the ownership of animal genetic resources in various gene-banks and the legal frameworks for the databanks (Figure 11.3 and 11.4).

11.18 HARNESSING IP-LINKED TECHNICAL OPPORTUNITIES IN AGRICULTURE

Trademarks should be used for brand development in agriculture. Genes and gene sequences, amino acid sequences, antibodies, etc., should be protected by copyrights until there is opportunity to patent and commercialize these products. Protection of IPR in all cases should be essentially linked to commercialization, sharing of royalty and other benefits. Appreciating that the agricultural research community should create/innovate, protect, and commercialize their new technologies on continuous and incremental basis, other important responsibilities, like sustainable development, empowerment of weak farmers, and protection of their traditional resources and knowledge should also be prompted on high priority. Quick action should be taken to record and document farmers' varieties in the country as available over space and time. Biotechnological advances should be integrated with genetic resource management where feasible to identify

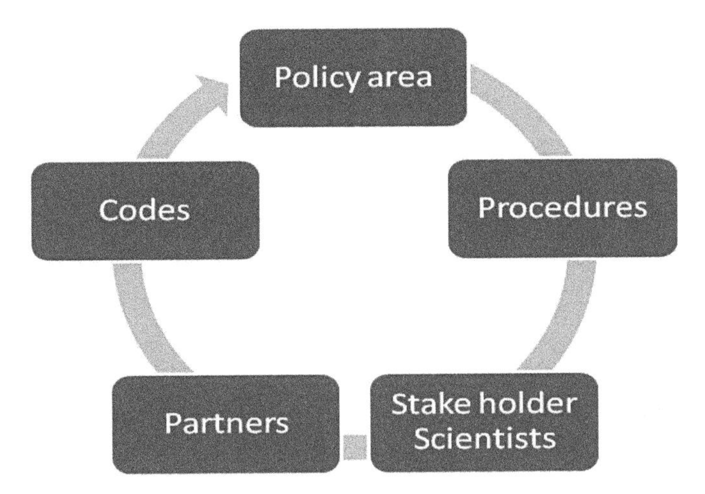

FIGURE 11.3 The figure shows the integration of stakeholders through policy reform in IPR.

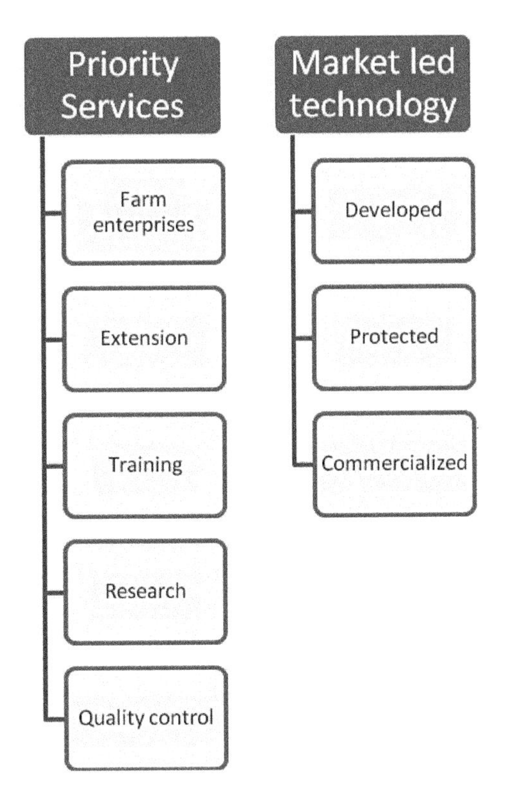

FIGURE 11.4 Different policy areas to be strengthening in agriculture.

copyright and document unique genes or gene sequences. The germplasm registration is in practice, and specifications and guidelines should also be developed for breed registration of livestock. Certain percentage of the research budget in agriculture should be allocated to protect the public-sector R&D for IPR portfolio management, and technology development and mobilization in agriculture. Certain technologies are important for food security, but significant avenues do not exist for IPR protection and commercialization. As the IP protection is likely to be more stringent, agricultural markets should be constantly monitored. Funding schemes should be encouraged to develop research links between profit-making and non-profit-making research institutions and to bridges between the use of propriety and public domain resources and technology.

11.19 LINKAGES AND CO-OPERATION

Mutually supported technologies should be encouraged, in public-public, public-private, or private-private partnerships to address high proportionate initial costs and risks. Partnerships should be encouraged in exploring the new tools of applied genomics to improve the biological systems. The minimal codes of procedures should be developed and applied in different key areas of partnership. Confidence building should be accelerated in cross-sectoral partnerships. Opportunities should be provided for frequent interaction among the agricultural scientists, research institutions, agricultural industrial sector and entrepreneurs. The private sector should also complement the basic and strategic research by the public sector through appropriate funding. Voluntary legal advice may be provided in partnership deals of strategic importance to enhance competitiveness of Indian agriculture. A common platform should be provided to seek assistance from the attorneys and lawyers having reasonable agricultural background. Besides, outsourcing for legal advice in order to competently address the techno-legal area of IPR in agriculture, should appoint law officers in their IPR Cells. The jurisdictional limits in respect of the application of IPR laws and the situations concerning enforcement and discipline, and control of agribusiness abroad should be addressed by all concerned in a national spirit. Agencies like APEDA, FICCI, and CII should earmark resources and funds to meet the contingent needs for transnational IPR cases involving the Indian agricultural and to provide emergent support.

11.20 CONCLUSIONS

The classical IPR in agriculture are patents, particularly on biotechnological inventions, plant breeders' rights, trademarks, and GIs. The Trade secrets and the protection of undisclosed test data are considered to be part of IPRs. The Farmers' rights and community IPRs are the forms of intellectual property at the stage of initial conceptualization. India is a member of the WTO and is therefore, obliged to implement the TRIPS Agreement within the time limits. Most of the TRIPS obligations on the IPRs, including strong process patents for biotechnological inventions, and it is only for product patents on microorganisms.

Recently, India has proposed the enactment of a biodiversity law. The government of India wants to encourage investment by private seed companies, as evidenced from its policies since the mid-'80'S, plant breeders' rights would help in giving incentives for private research. The deployment of provisions on compulsory licensing and government use and the recognition of the mutual inter-dependence between public sector and private sector research efforts, may resolve the dilemma of incentives for generation of technologies. The Consultative Group on International Agricultural Research (CGIAR) and the International Agricultural Research Centers (IARCs) plays a constructive role in the two-way transfer of technologies between the (National Agricultural Research Systems (NARS) and private sector seed companies. Several modalities have already been envisaged such as Material Transfer Agreements, licensing or cross-licensing, joint ventures or private funding of basic research in the public sector.

The legislative exercises on amending the Patents Act, 1970, particularly on the patenting of biotechnological inventions should be made more transparent, with the involvement of all stakeholders such as agricultural scientists, farmer groups, private sector seed companies, lawyers, experts, and NGO activists. Similar exercises are required to implement the TRIPs provisions on undisclosed information. This would require the conduct of workshops and the setting up of drafting committees as well as the building up of mutual trust.

Recently, there is vocal demand made by sections of the media to introduce *sui generis* legislation for the domestic protection of GIs such as basmati rice. In addition, a conscious effort needs to be made to build up the brand equity of Indian markets. India should seek to conclude bilateral agreements with interested WTO members within the framework of the TRIPS Agreement. Institution should improve patent literacy among scientists. It is time to enact the required legislation and the implementing rules and regulations, allowed under TRIPS. This book chapter does not make any claim to a complete or exhaustive list of all that needs to be done for IPRs in agriculture in India. It merely emphasizes the immensity of the tasks that remain to be done in the light of the sharp differences of opinion amongst the stakeholders. There is an urgent need in different organizations to gear up to contribute to this exercise in a meaningful way.

KEYWORDS

- **agricultural education**
- **agricultural technology**
- **IP laws**
- **IP licensing**

REFERENCE

IPR knowledgebase: IPR overview, http://www.rkdewan.com/iprOverviewMain.jsp.

INDEX

C

D

J

K

T

Y

Z